*Economics—Mathematical Politics
or Science of Diminishing Returns?*

Science and Its Conceptual Foundations
David L. Hull, Editor

ECONOMICS—
Mathematical Politics or Science of Diminishing Returns?

ALEXANDER ROSENBERG

THE UNIVERSITY OF CHICAGO PRESS
Chicago and London

ALEXANDER ROSENBERG is professor of
philosophy at the University of California, Riverside. He
is also the author of *Microeconomic Laws: A Philosophical
Analysis, Sociobiology and the Preemption of Social Science,
Hume and the Problem of Causation* (with T. L. Beauchamp),
The Structure of Biological Science, and *Philosophy of Social
Science.*

The University of Chicago Press, Chicago 60637
The University of Chicago Press, Ltd., London
© 1992 by The University of Chicago
All rights reserved. Published 1992
Printed in the United States of America

00 99 98 97 96 95 94 93 92   5 4 3 2 1

ISBN 0-226-72723-8 (cloth)

Library of Congress Cataloging-in-Publication Data

Rosenberg, Alexander, 1946–
    Economics : mathematical politics or science of
diminishing returns? / Alexander Rosenberg.
        p.   c.m. — (Science and its conceptual foundations)
    Includes bibliographical references (p.   ) and index.
    1. Economics—Philosophy.   I. Title.   II. Series.
HB72.R66   1992
330—dc20   92-140
                                                                 CIP

⊚ The paper used in this publication meets the
minimum requirements of the American National
Standard for Information Sciences—Permanence of
Paper for Printed Library Materials, ANSI Z39.48-1984.

For five friends:
Bruce Bueno de Mesquita, Tom Beauchamp,
Bob Martin, Peter Van Inwagen,
Carol Tomlinson-Keasey
With admiration and gratitude

CONTENTS

CHAPTER 4

Neoclassical Economics as a Research Program   87

CHAPTER 5

Economics and Intentional Psychology   112

CHAPTER 6

Could Economics Be a Biological Science?   152

CHAPTER 7

Why General Equilibrium Theory?   200

CHAPTER 8

Is Economic Theory Mathematics?   228

PREFACE

Economic theory is a perplexing subject. Though I have spent the better part of my academic career thinking about its aims and methods, I have never been confident that I or anyone else for that matter really understands its cognitive status. Partly no doubt this is because everyone's understanding of the cognitive status of most intellectual disciplines has been subject to great disturbances over the last two decades or more. Even at the time I first began thinking and writing about the problems that economics presents for the philosophy of science, in the late sixties, the conceptual framework within which scientific disciplines were assessed was coming under serious question. Since then matters have moved so far in the philosophy, history, sociology, and psychology of science that the very notion of "cognitive status" has gone into eclipse.

Over the same period that philosophy's sense of certainty about what science is has disappeared, the philosophy of economics has emerged as a growth industry, and many people are now confident they understand the aims and methods of the discipline. When I published my first paper on Milton Friedman's famous "methodology for positive economics" in 1972, there was nothing to be found from the pen of philosophers on it, save for an excellent assessment by Ernest Nagel. Other than a small number of papers by Samuelson, Simon, Koopmans, and Machlup, economists had been equally reluctant to turn their attention to the problems Friedman's important essay dealt with. It is true that there had been a volume at the turn of the century by Keynes *père* and a period of methodological reflection among British economists in the thirties, of whom

Robbins and Hutchison were the principal figures—but after that almost nothing for thirty years.

Yet by the beginning of the 1990s, the philosophy of economics had become a recognized subdiscipline within philosophy. The history and sociology of the discipline have grown apace as well, in part because the philosophy of science has increasingly found mutual interests with the history and the sociology of all the sciences.

By now the philosophy of economics has acquired all the impedimenta of an academic discipline, several dozen monographs or more, a couple of anthologies of canonical papers, textbooks by household names, a journal, an international society or two, and an "invisible college" of scholars exchanging papers before publication, airing their differences in a half dozen journals of varying degrees of specialization.

As those who have read papers of mine over the last twenty years will testify, my own views about the nature of economic theory have remained permanently unsettled, partly as a result of all this activity. Since the appearance of my first book, *Microeconomic Laws: A Philosophical Analysis*, about the only thing that has remained fixed about my approach to this subject are the questions with which I started: questions about the cognitive status of economic theory, that is, questions about whether that theory is to be understood and assessed for adequacy along the lines of theories in the sciences—physical, biological, behavioral. Over the period since I wrote *Microeconomic Laws*, these questions about cognitive significance have not been popular. But I have stuck with them, largely because the answers can have significant ramifications for public policy, for our hopes to improve its intended effects and mitigate its unintended ones. If economics is a science, if its theories are composed of improvable regularities about human behavior and its aggregate consequences, then its bearing on policy is as evident as the study of physiology is to human health. If not, confidence reposed in it will be repaid by frustration at best. Without some assurance about the cognitive status of the theory, there is no basis for confidence in it. Now the trouble with this question about the cognitive status of economic theory is that in the philosophy of science there is no longer any conviction about the existence of a litmus test that will indicate the cognitive status of a theory. Though some economists still pay lip service to Popper's litmus test of falsifiability, most students of the philosophy of economics recognize that (*a*) economic theory is not in fact falsifiable, and (*b*) falsifiability is unacceptable as a test for the scientific respectability of a theory. But no adequate substitute litmus test has been found. The absence of such a test has led some students of the subject to conclude that there is in fact no difference in the cognitive status of economics, physics, mathematics, literary criticism, or astrology, for that matter. Although I recognize that there is no litmus test here, I persist in believing that there are important differences between the cognitive status of these various enterprises. It's just very difficult to establish what these differences are.

When I wrote *Microeconomic Laws* I argued that there was no conceptual obstacle to microeconomic theory's status as a body of contingent laws about choice behavior, its causes and consequences. In the years since I wrote that book, I have not changed my mind on this score, but I have come to believe that this conclusion takes us only a very little way toward understanding the nature of economic theory. And so in a series of papers since the late seventies, I have tried to go further. The results have been in some respects indecisive; I have changed my mind about important issues more than once and have paid the price in criticisms for views I no longer found myself holding or had seriously qualified. But economic theory is a difficult enough subject to put in perspective that perhaps such changes of mind will be excused. At any rate the views I have held, including some I no longer hold, have served as useful targets for other writers in the field, who have employed them to at least identify positions against which they can clearly delineate their own understanding of economics.

Though my thinking has shifted repeatedly in the past, it does now seem to have found an area of at least local stability. That fact and my continuing conviction of the importance of the questions that initially stimulated my writings on this subject have emboldened me to write this successor to *Microeconomic Laws*.

Readers familiar with debates in the philosophy of economics will recognize the issues broached in the chapters that follow. Some may recognize in these chapters distant ancestors in papers I have written since the late seventies. Those coming fresh to the philosophy of economics may find a little stage setting useful.

Chapter 1 lays out what I believe to be the relation between a discipline like economics and the philosophy of science. I argue that a philosophy of science does to some extent constrain science and that such constraints are in fact stronger in the social sciences than in the natural ones. But the philosophical constraints are established by scientists as much as by philosophers. In fact the constraints I identify are distinguished from substantive theoretical claims only by their abstractness and generality. The chapter concludes by drawing some morals for economics from the existence of philosophical constraints. In particular I suggest that the most innocuous epistemological requirement on any science is that its theories can be improved in predictive precision. This does not mean that they must start out showing great predictive power, only that theories need to show improvements in predictive power, and scientific disciplines need to reflect an interest in securing such improvements.

Chapter 2 locates my view in a spectrum between two extremes: according to one of which, an economic theory stands or falls on the adequacy of its underlying philosophy of science, and according to the other, theories in two subjects are wholly independent of one another. The first of these views is traditionally associated with Marxian economic theories, which often emerge as seamlessly bound to Marxian philosophy. Its clearest elaboration in recent years is to be found in the work of Martin Hollis and Edward Nell. The chapter begins with a brief examination of their view. The second, now more fashionable, thesis, that philosophy is irrelevant to economic theory, follows from an approach to the nature of economic reasoning championed by Donald McCloskey. I spend most of chapter 2 attempting to undercut the force of McCloskey's "rhetorical" approach to economics, because I consider these views inimical to the health of economics as a discipline.

Chapter 3 defends the claim that economic theory has not satisfied the minimal requirement of predictive improvement or even shown much of an interest in securing it. The absence of predictive improvement or even an interest in it will be denied by most economists. And it may be viewed as impertinent if not impudent for a noneconomist to gainsay their claims. My strategy therefore in this chapter is to ring the admission of predictive weakness from defenders of neoclassical economies, like Milton Friedman, Paul Samuelson, and Hal Varian, and to cite equally distinguished economists, like Vassily Leontief, who have argued that economic theory remains predictively weak. The chapter includes an extended example from the recent history of economic theory that reflects the discipline's indifference to predictive improvement, in contrast to its desire for formal tractability.

Some writers on the philosophy of economics have recognized the insularity of its theory from predictive improvement and the indifference of its leading figures to such improvement. But they have argued that this is unexceptional in division of labor of a scientific research program. Following Imre Lakatos, a number of economists have attempted to justify the insularity of economic theory by clothing it in the garb of the hard core of such a program, insulated from the hurly-burly of empirical testing. In chapter 4 I examine this way of responding to the admission that economic theory is predictively weak. I argue that Lakatos's approach is unavailing as an explanation or justification for the weakness of economic theory. Indeed, the methodology of scientific research programs is an unstable position, ready to collapse into the rhetorical approach to economics, as the intellectual shifts of E. Roy Weintraub—one of its most able defenders—show.

In chapter 5 I begin to develop my diagnosis of why economic theory remains predictively unimproved and unimprovable. Like theories in other social sciences, economic theory explains human behavior as the result of desires and beliefs, though it labels them preferences and expectations. Like these other theories it is hobbled by the recalcitrance of these variables to measurement independent of their effects in actual choice. An excursion into the philosophy of psychology is needed to see that the recalcitrance to measurement of beliefs and desires is unavoidable and beyond a certain point irreduc-

ible. Chapter 5 provides the excursion which seeks to explain
this unavoidable irreducibility.

Chapter 6 deepens and illustrates the arguments of chapter 5
by considering Gary Becker's well-known attempt to circumvent
problems of untestability that daunt the field's approach to the
nature of preferences. Becker's theoretical innovation is to assume that preferences are stable and uniform across agents. I
argue that properly understood, Becker's approach can in principle solve half the problem of predictive unimprovability, but
only in principle and only one half of it. The half of the intentionality problem left untouched by Becker's proposal is the
problem of a disaggregated extensional account of information.
About the solution to this problem I hold no hope.

Economic theory's commitment to equilibrium explanations
combines with its intentionality to generate its predictive unimprovability. In chapter 7 I examine several arguments advanced
in favor of the equilibrium approaches to economic theorizing.
I find none satisfactory and finally offer one of my own. Although it explains and perhaps even justifies the role that equilibrium has come to be granted in economic theorizing, it does
nothing to improve the chances of the theory's predictive improvement. The conclusion of chapter 7 leads, in chapter 8, to a
reexamination of the claim I made some years ago that economic theory ought best to be viewed as a branch of applied
mathematics. The view that economic theory is best understood
as part and parcel of the enterprise of mathematics attracted a
good deal of hostile attention when I first advanced it. Here I
reexamine the thesis, hoping that the seven chapters that pre-
cede this one will have provided something of an argument for
this view.

As the subject of the philosophy of economics grew over the
last two decades I have come to be indebted to an increasing
number of philosophers and economists. These students of the
subject understand both philosophy and economics far better
than I did when I began, and better than I would now, without
their help. In many areas their understanding still vastly exceeds my own. Among them the most important have been philosophers like Daniel Hausman, Alan Nelson, Alan Gibbard,
Phillipe Mongin, David Braybrooke, Jules Coleman, Margaret

Shabas (a historian as well as a philosopher), Jules Sensat, Gilles Gaston Granger, Martin Hollis, Alfred McKay, Christina Bicchieri, Jon Elster, Harold Kincaid, and David Pappineau. The economists to whom I am most deeply indebted include Roy Weintraub, Bruce Caldwell, Wade Hands, Niel De Marchi, Hal Varian, Mark Blaug, Amartya Sen, Richard Cyert, James Buchanan, and Herbert Simon.

As will be evident, none of these economists and philosophers can be held to agree with anything I say. In fact, I take issue with many of them by name in the pages to follow. But each will recognize where and how he or she has influenced my thinking for the better. So will David Hull, Susan Abrams, and Michael McPherson.

This work was brought to completion with grants from the National Science Foundation, for which I am grateful.

1

WHAT DOES THE PHILOSOPHY OF SCIENCE HAVE TO DO WITH ECONOMICS?

Why should economists care about philosophy? In *Micro-economic Laws*[1] I tried to explain to philosophers of science why we should care about economics. Answers to this question, whether my own or others, seem to have been satisfactory, for few disciplines have more fully claimed the attention of philosophers over the last two decades than economics. More frequently students of the philosophy of economics have met with the question, Why should economists bother about philosophy of science?

There are to be sure some economists who are entirely satisfied on the point. They hold that economic theory rests on philosophical foundations. In order for the discipline to advance, the foundations must be secure. For them philosophical inquiries are foundational ones. These economists are a distinct, though influential, minority, and I examine a sample of their views in chapter 2. A more fashionable view nowadays answers attempts to pull the rug out from under the question, Why should economists care about the philosophy of science? by holding that the philosophy of science is a nondiscipline superseded in the passing of "modernism"—a now-defunct intellectual fashion associated with Descartes. I explore this view as well in chapter 2.

Between these two views are the positions of most economists: though they require to be shown why they should care about the philosophy of science, they are open to some positive reason to do so. In this chapter I hope to provide that reason. Providing a reason for economists to take seriously the philoso-

1. Pittsburgh: University of Pittsburgh Press, 1976.

1

phy of science is important for the argument of this book. Because I will make claims hereafter from which economists will dissent, if not stigmatize, as beyond the qualifications of a mere philosopher, it is important to establish the claims of the philosophy of science on the attention of economists.

My argument begins with the bearing of the philosophy of science on all the sciences and then narrows to its relevance for social science and economics in particular. The conclusion to which I come is the presumptuous one that the philosophy of science has after all a prescriptive role to play in adjudicating and assessing the promise and performance of scientific theories. In this respect it shares a common feature with a now long-defunct tradition of logical positivism. The only mitigation I offer for the presumptuousness of my conclusion is that if I am correct, then all scientists are philosophers of science; or at least all take sides, willy-nilly, on issues in the philosophy of science, and the sides they take influence the direction of their "substantive" work.

If the philosophy of science has a prescriptive role to play with respect to all the sciences, including economics, then economists have the best reason in the world for paying attention to the philosophy of science: it knows what is good for their discipline.

Many of the books and papers in the philosophy of social sciences, and almost all of the writing in the philosophy of economics, begin with obituaries for logical positivism. Most hold that its death has freed the sciences from the straitjacket of a narrow-minded empiricism and encouraged a thousand methodological and theoretical flowers to bloom, especially in the social sciences. The morals that have been extracted from the eclipse of positivism seem to me to be profoundly wrong. Once we see what these morals should have been, the prescriptive force of philosophy for the sciences, and in particular for economics, will become evident.

THE REAL RELATION BETWEEN PHILOSOPHY AND SCIENCE

The intellectual core of logical positivism is found in its attempt to characterize exactly the notion that claims must be empirically testable if they are to constitute real accretions to the fallible store of reasonable scientific belief. In the terms of the positivists, scientific statements had to be cognitively significant, and positivists spent the better part of three decades attempting to devise a litmus test for cognitive significance. The program of positivism came to a halt when the logical positivists and their students gave up their attempts to formulate a criterion of cognitive significance. The last serious effort to propound such a criterion was Rudolph Carnap's "The Methodological Character of Theoretical Concepts" in volume 1 of the *Minnesota Studies in the Philosophy of Science*.[2] Beyond the technical objections to Carnap's criterion, it came increasingly to be recognized that the insurmountable problem for a principle of cognitive significance is the absence of any demarcation between scientific and nonscientific discourse. There is no litmus test for empirical or factual statements to distinguish them from either formal or analytical ones or theoretical and speculative ones.

Criteria of cognitive meaning had, among positivists, the function of distinguishing between scientific discourse and meaningless metaphysics, between a posteriori claims and a priori ones, and between prescriptive or normative claims and positive or descriptive ones. But no adequate criterion could ever be formulated, because the smallest unit of empirical meaning is a whole theory, that is, a body of statements that cannot be segregated into neat divisions of factual versus nonfactual, analytic versus synthetic, observational versus theoretical.

We cannot distinguish between the "cognitive significance" of factual claims about observations and theoretical claims that transcend observation, or between either of them and the laws of logic or the rules of methodology or the conclusions of metaphysics. These distinctions cannot be drawn because there is no such thing as "cognitive significance" or because statements of all these different types are equally significant on any acceptable criterion. This is not to say that there are no distinctions between these different sorts of statements. There are clear cases of each of them, and some differ from others syntactically.

2. Edited by Herbert Feigl and Michael Scriven (Minneapolis: University of Minnesota Press, 1956), pp. 38–76.

Thus, there may be no one clear-cut distinction between observable and nonobservable statements, but there are clear cases of each of them; similarly we can distinguish methodological prescriptions from factual claims, if only because of their grammatical form as imperatives. But the cognitive status of all these different sorts of claims is the same: each of them is underwritten or undermined by the same sort of considerations. This identity in cognitive significance is the lesson of the positivists' valiant effort to construct a criterion of cognitive significance. The history of this effort showed that the only criterion general enough to determine cognitive significance of empirical meaning will have to focus on the form of scientific language and the epistemic warrant of claims expressed in it. But the linguistic form of scientific claims cannot be distinguished firmly enough from their factual content to do the job; and there seems to be no class of statements in any language, natural or artificial, with claim to exclusive or maximal epistemic warrant.

To see how this denouement of the recent history of the philosophy of science reveals the actual relationship of the philosophy of science to the methods of the particular sciences, we need to be clear on what these methods come to. Presumably, the methods of science are expressed in rules. Consider the following examples of methodological rules in physical and social science:

Always calibrate a pH meter against distilled water.

The explanation of why this rule should be followed is so transparent to laboratory scientists that it usually goes unmentioned. Nevertheless, there are explanations available, ones whose full details vary with the complexity of the pH meter's design, for instance. No one would normally suppose that the philosophy of science has much to say about why this rule should be followed.

Chemical equations should be stoichiometrically balanced.

Considerable effort is devoted to explaining and justifying this rule in chemistry. A reasonably full explanation will advert to principles of the conservation of mass and electrical charge, as well as to elementary versions of atomic theory. Some of the story here will also be part of a complete explanation of the pH rule as well. But again, it is doubtful that philosophical considerations bear on the explanation of why this rule should be employed.

> When possible, employ double-blind experimental procedures.

Double-blinding is an important methodological rule in pharmacology and much social psychology. Unlike the previous rule, it is not explained and justified by any very detailed theory of "experimenter effects" or "placebo effects." It nevertheless does have considerable justification. Its basis is in important experiments that reveal a causal relationship between experimental expectations and outcomes. We need a good theory to explain and justify more fully this rule, but in its absence we still have considerable evidence that it should be followed. And this justification in turn rests on following another methodological rule, one common to the natural and social sciences:

> Same cause, same effect.

This principle takes a stand on the relevance of regularities and laws to the identification of causes and their effects. It and associated principles were long held to be a priori knowledge that submitted of metaphysical justification as necessary truths. They continue to be viewed as quite different from the first three examples, as somehow not on the same scale as the methodological principles that govern day-to-day natural science. On the other hand, among social scientists this rule seems to have greater salience, both as a subject for controversy itself and as a substantive research technique.

Here is a methodological rule from economics:

> When markets do not clear, search for impediments to price flexibility.

Unlike the double-blind rule, this rule rests almost entirely on theory instead of experiment: its explanation and justification are to be found in microeconomic theory, and the issue of whether it is well justified by any empirical evidence, as the double-blind rule is, is highly controversial. Much of the rest of this book focuses on the warrant for adopting methodological rules in economics like this one.

Frame hypotheses and test them. Reject falsified ones. Tentatively accept confirmed ones.

This rule is well known to philosophers under the name hypothetico-deductivism. Many have accepted it as the sole method of scientific advance. But the fact that it was widely rejected for long periods in the history of science belies this assurance.[3] Surely, any explanation and justification for following this rule must be largely philosophical. Moreover, the relation of such explanations to explanations of more substantive methodological principles is quite unclear.

Be logical: do not commit fallacies in reasoning.

The justification of this rule long seemed unnecessary. Yet the rule has no content unless we know what the principles of logic are. The conviction that we know what these principles are has been shaken by, for example, the alleged incompatibility of quantum mechanics and the rules of two-valued logic. This alleged incompatibility shows at least that the laws of logic may be explained and underwritten in ways no different from the ways we establish more narrow methodological principles.

Despite appearances, our examples are not categorical imperatives, but hypothetical ones. They are not prescriptions that bind all rational agents, only those with certain aims. In effect they are all conditional or hypothetical imperatives with suppressed antecedents. Each should be read as follows: "In order to attain the goals of science, . . ." with the rules exemplified above replacing the ellipses (subject to suitable grammatical reformulation). The hypothetical imperatives in which methodological rules are embedded do not specify the goals of science. Still less do they commit us to the existence of a small number of attainable or recognizable goals on which all parties can agree. Indeed, some will hold that science as an institution does not have goals, or at least none that really deter-

---

3. See, e.g., Larry Laudan, *Science and Hypothesis* (Dordrecht: Reidel, 1981), for a discussion of the checkered history of the acceptance of this rule.

mine scientific practice. Rather, individual scientists have goals, and these may vary from individual to individual. But even such writers allow that all scientists at least pay lip service to the doctrine that the goal of science is knowledge. For purposes of my argument, this concession may be enough.

Every long-lived methodological rule has a history of successes, circumstances in which it was employed and should have been. And all but the ones currently in force have cases of failure, circumstances in which they were employed and should not have been. In the former cases the application of these rules sometimes leads to scientific advance—whether theoretical, empirical, or practical; in the latter cases it does not. Thus, for instance, exploiting these rules sometimes leads to new experimental results that affect the credibility of available theory, or provokes new theory; sometimes it results in technological advance or improved powers to predict, control, and prevent occurrences, and so on. Sometimes following these rules produces sterility, frustration, and even regression. Both the particular successes and the failures need explanation and have causes. What determines the successful application of a methodological rule? What distinguishes those circumstances in which following a rule enables the scientists to attain their goals from those in which it does not?

In addition to their implicit form as hypothetical imperatives, methodological rules also embody certain distinctive descriptive claims about nature, about the way the world is. Thus, for example, the pH-calibration rule reflects the supposition that the pH of distilled water is a constant. Of course the rule has many other assumptions, implications, and entailments. But this one is particularly immediate and independent of many of these other conditions and ramifications of the rule. These "immediate assumptions," as I shall call them, are closely related to the explanation of a rule's success. The relation is varied, but in the simplest case, it has roughly the following character. The immediate assumptions to which the rule commits us are true, and their truth is part of an (actual or possible) explanation of the rule's success. That is, the assumption describes facts that causally determine in the circumstances the success of the rule, and these circumstances would be cited in an explanation of why the rule works. For example, the injunction to calibrate a pH meter against distilled water is successful (i.e., enables chemists to determine accurately the pH of various solutions), in part because its assumption, that the pH of distilled water is a constant, is a causally necessary condition for the rule's success that actually obtains in the circumstances of the rule's application.

In most cases, however, the situation is more complex, and the relation between a methodological rule, its immediate assumptions, and the causal determinants of its success are more indirect. For example, consider the rule expressed in the maxim "same cause, same effect." Among this rule's assumptions is obviously the supposition that events of the same kind have effects of the same kind. Now, beyond a certain level of precision in microphysics, this statement is false and cannot be part of the actual causal conditions that determine its success in application to macrophysical phenomena. So, the explanation of success in this case must appeal to considerations that the rule does not immediately assume. In particular, the explanation relies on the fact that quantum-mechanical phenomena asymptotically approach determinism at comparatively low levels of aggregation of the microphysical constituents of matter. The asymptotic approach to determinism explains why the assumption that events of the same kind have effects of the same kind is a good approximation to the quantum-mechanical description of macroscopic phenomenon, and thereby explains why the methodological rule "same cause, same effect" is often a successful one. Moreover, these same considerations will explain why following the rule fails in, say, high-energy physics.

So, every methodological rule is thus part of a wider hypothetical imperative of the form:

In order to attain the goals of science, employ rule R.

And each rule embodies immediate assumptions about the world which would figure in any explanation of its success or failure. Call them t, for "theory," and symbolize the rule that embodies t as $R[t]$ to reflect the dependence of the rule on the theory. Now, the justification for employing the hypothetical imperative $R[t]$ is that acting in accord with its consequent in fact does help attain the goal referred to by its antecedent. And in the simplest case this fact is explained in part by t. t will also be part of the explanation of why $R[t]$ succeeds, when it does so. Naturally, the best and most complete explanation for the success of many different methodological rules will share many elements in common, but the explanation for the success of $R[t]$ will differ from that of the explanation of another rule $R[t']$], at least because the former will include t whereas the latter will include t'.

Even where $R[t]$ is successful while t is false, t will still have a role in the explanation of $R[t]$'s success. Why? Because if $R[t]$ is successful, then there is some theory or other, known or unknown, that is part of the explanation of $R[t]$'s success. It does this because it explains t, shows why t is a reasonable though false theory to embrace under the circumstances in which $R[t]$ is employed, or shows why those who embraced t were right or right enough by accident to explain $R[t]$'s success.

We may exploit this relation between a methodological rule and the suppositions it assumes to distinguish between the scientific justification of a rule and the social, historical, psychological, or indeed the rhetorical explanations why scientists employ it. When using a rule results in a scientific success, the part of the explanation of why it was employed that is not shared with explanations of the successes of any other rule is what justifies its use. Insofar as a general theory of scientific behavior is possible, it should be able to explain a variety of different scientists' choices, beliefs, and so on, in several different disciplines. That is, it should be the common element in many explanations of why scientists employ a variety of methodological rules. Thus, we may expect that the explanation of why social scientists employ the method of significance testing will appeal to the same social forces that help explain why pharmacologists employ double-blind procedures. For example, journal editors in their fields demand such methods and the scientists wish their research to be published.

But some elements in the explanation of why a scientist uses a particular rule can be expected to be distinctive and restricted in their relevance to that rule. Whatever is peculiar to such an explanation of why that rule is used will be part of the scientific justification of the rule, and not merely part of the explanation of why a particular scientist used it. Thus, the placebo effect (along with the Rosenthal effect) is part of the scientific explanation of why pharmacologists employ double-blind techniques and is not part of the explanation of why sociologists employ significance tests. The theory that there are placebo and other experimenter effects is the immediate assumption of the double-blind rule. It is the t to this $R[t]$. The set of propositions peculiar to the explanations of why rules are employed successfully will be all or most of accepted scientific theory. These propositions not only help explain why a rule has been used but are also part of the justification of employing it, by contrast with mere causes—social, psychological, ideological—for employing it.

Thus, advocacy of a methodology cannot be neutral on the truth or falsity of the theory it presumes. And every methodology that claims success presupposes a theory about why it is successful, a theory not about the users of the method that purports to cite the causes of why they use it, but a theory about the subject matter that explains why its use is or will be successful. There is an important asymmetry between methodological rules and theories. The success of any methodological rule $R[t]$ is to be explained (in part) by t or some other theory, t', known or unknown, that bears an explanatory relation to t. Similarly, the failure of any methodological rule is to be explained by some theory, perhaps even the same theory that explains its successes. On the other hand, theories cannot be explained by methodological rules: the success of a methodological rule is not among the causal determinants of the truth of any theory. Naturally, the appropriate use of a methodological rule can be part of the story of why or how we have hit upon a theory, but it is no part of the truth conditions of the theory or of the verisimilitude conditions of it. Similarly, the rules employed in the disconfirmation of a theory are no part of what makes the theory false, though they may be part of the cause of our disconfirming the theory.

The asymmetry of theories to rules is also the basis of the prescriptive roles of the former. Not only do theories explain the successes and failures of rules that assume them, they also prescribe these rules. Or at least, they do so when harnessed with the goals of science. The goals of science, together with particular theories, give particular methodologies their force. The prescriptive force of a theory derives from the fact that it identifies the methods that will lead to the attainment of the goals of science. Some illustrations of this relation between goals, theories, and rules are given below, with special reference to philosophical theories and economic methodological rules. For the moment it is important to see that philosophical theories function exactly as scientific ones do in constraining rules, given scientific goals.

The priority of theories over rules extends from the most restricted and particular methodological principle to the most general and unconditional among them. This priority will include those rules so broad and overarching that the theories which are prior to them are identifiably philosophical. Therefore, in the absence of a demarcation principle, philosophy's bearing on methodology is indistinguishable from that of the substantive particular sciences. Behind every methodological rule there stands a theory, a body of statements, explaining its successes, and for every rule that fails, a body of statements explaining it failure. What will these theories contain? The material findings of science, naturally, but philosophical, logical, epistemological, and metaphysical theses will be included as well, for the truth or falsity of such theses will be part and parcel of the explanations for the failures and/or successes of these rules. Their relevance to such explanations cannot be ruled out on the grounds that such philosophical claims are necessary truths, conventions, meaningless metaphysical speculation, rational reconstructions, or reports of usage, ordinary or extraordinary.

If theory adjudicates the rules of science, then so does philosophy. In the absence of demarcation, philosophy is just very general, very abstract science and has the same kind of prescriptive force for the practice of science as any substantive scientific theory. Because of its generality and abstractness, it will have less detailed bearing on day-to-day science than, say, prescriptions about the calibration of pH meters, but it must have the same kind of bearing.

In fact, for some methodological prescriptions of the highest generality and applicability, we should expect only philosophical explanations of success and failure. Additionally, it is reasonable to expect philosophy to have the most prescriptive bearing in disciplines less theoretically well-developed than, say, chemistry. In the absence of narrower theories to identify and explain discipline-specific rules, there will not be any very specific methodological rules sanctioned in such disciplines. The only source of explicit methodological rules for these disciplines will be theories in already-established subjects. But the appropriateness of such rules in a new and as yet undeveloped discipline can be affirmed only by philosophical considerations, for these rules will perforce be extremely general if they really are applicable across a wide range of subjects. But this means that philosophy will have more relevance for theoretically underdeveloped disciplines, like the social sciences, or even economics. This does not mean that any particular philosophy of science must eventuate in just one body of economic theory or method. It means that explanations for the appropriateness or inappropriateness of certain methods and arguments for the fruitfulness of theoretical approaches in the less-developed disciplines will inevitably be philosophical.

THE MUTUAL CONSTRAINTS OF GOALS, THEORIES, AND RULES

The goals of science, together with available theory, constrain its methodological rules. We can illustrate this relation at several levels of generality. At a relatively low level, chemists whose goal is predictive success and who adopt the standard theory of the nature of acids and bases are committed to the rule about calibrating pH meters with distilled water. Philosophical theories enter at a much higher level of generality. Thus, psychologists whose scientific goals include predictive success and whose methodological rules restrict inquiry to observable behavior only are committed to a philosophical theory that casts doubt on the causal role of intentional factors like preferences and expectations. Economists with the same goal and a commitment to neoclassical methodological rules are committed to the denial of the philosophical theories the psychological behaviorists must embrace. Economists who repudiate predictive success as a desirable or attainable scientific goal and opt for the same methodological rules are committed to a disjunction of philosophical theories. That is, they must reject either the notion that expectations and preferences are causes or the theory that causes operate in accordance with regularities that enable us to predict their effects, or they must endorse some philosophical theory or other to the effect that human behavior is in principle unpredictable. Economists who repudiate prediction as a goal of economic theory are committed to one or another such philosophical thesis, unless they hold that prediction is merely a practical and not a theoretical impossibility. But then, one needs an explanation for this impossibility.

Part of the task of the philosopher of science is to reconcile alternative epistemologies with actual scientific theories and the methodological rules scientists employ, for these alternative epistemologies are in effect specifications of the goals of science. By setting out definitions of knowledge, they prescribe the conditions that must be met to attain it. An empiricist epistemology erects predictive success as a goal of science, just because it asserts that beliefs require empirical confirmation in order to be certified as knowledge. By contrast, a rationalist epistemology prescribes certification of beliefs as necessary truths if they are to count as knowledge and thus imposes a different goal for science, or at least for economics.

Differing philosophies of science reconcile goals, theories, and methods by, so to speak, holding one or two of these three variables constant and adjusting the others to suit. A stipulation about the epistemic goals of science combined with a certain body of theories about science will constrain scientific method. Similarly, given the methodological rules and the accepted theories about science, we can extrapolate its goals. Or again, given its goals and its rules, a considerable amount can be said about the kind of theories that characterize a discipline. For example, the radical empiricist adopts an epistemology which restricts attainable knowledge to what can be observed only. Accordingly, such an empiricist is obliged to offer a special interpretation of theories that appear to transcend observation, and of rules that do so as well. He or she attempts to construct an interpretation of theories and rules that does them justice while making predictive adequacy science's preeminent goal. By contrast, the scientific realist holds constant the scientist's own interpretation of theories and methods and argues that they together dictate an epistemology which makes knowledge beyond observation possible.

Anyone setting out to examine a particular discipline and its rules, like economics, for instance, can proceed in a number of different ways. For instance, one may work backward from an identification of the methodological rules and the theories that justify them to an implicit identification of what the goals of economics are. Working backward from the accepted rules and theories of a particular science is certainly the most popular strategy in the philosophy of economics at present; it certainly seems to be the favored strategy of Donald McCloskey, whose views I examine in chapter 2. Alternatively, one may identify a set of goals as appropriate ones for any science and then assess the theories and rules of economics in the light of these goals. Starting with a set of goals was the strategy of *Microeconomic Laws*. The first method has the virtue of respecting the actual practice of economists and their own interpretation of their theories, but it has the vice of remaining silent on the degree to which economics attains scientific goals independently established. The second method has the virtue of taking these goals attained in other disciplines seriously but the potential vice of stigmatizing economic theory and method as inappropriate means to attain these goals. Of course the ideal situation is one in which whatever direction we move, we come up with the same answer; from prior goals we conclude that the actual character of economic theory and method is suitable to their attainment, or from the assumption that economic theories and methods are appropriate to the goals of economics, we conclude that these goals are the same as those of other sciences. Alas, all parties at present seem to agree that such harmony is not in the cards. Either there is something wrong with economics or our interpretation of it, in the light of the goals of science, or there is nothing wrong with economics and something seriously wrong with our identification of the goals of this discipline.

Which is it? The answer to this question can be settled only by identifying the goals of science. Now one can contend that there is no single or small number of "goals" we can reasonably assign to all the sciences or that differing groups of sciences—the physical, the biological, the social, the behavioral—have differing goals, or indeed, it may be claimed that each science has its own distinctive goals, or finally, perhaps sciences have no goals, only scientists have goals. This is doubtless the topic for a grand debate among philosophers of science. But in fact the multiplication of scientific goals seems unwarranted, for surely it is undeniable that all the sciences have for their ultimate goal knowledge. This claim is less vacuous than it appears. Because once we agree on what will count as a practical mark or a nontrivial necessary condition for knowledge, we have fixed at least the proximate goals of science: the attainment of those marks and necessary conditions. These marks of knowledge will of course figure in the definition an epistemology provides for knowledge.

The debate about whether all science or each science has one or more goals is in fact a debate about what each discipline singles out as a mark or criterion or necessary condition of knowledge. Of course, if what counts as knowledge differs among the disciplines, if the criteria for certifying knowledge in each discipline differ, then each will have different proximate goals. If there really are different kinds of knowledge, then each kind sets a different and potentially incompatible proximate goal from the others. The definition of each of the different kinds of knowledge we countenance sets forth a different set of necessary conditions or criteria for a claim's counting as knowledge of that type. Thus, if it is empirical knowledge we seek, we shall have to submit our claims to observational tests, at least indirectly. If on the other hand what counts as knowledge is what provides interpretative understanding, then the test for knowledge will be a hermeneutical one: coherence with our own intentional scheme. And if what counts as knowledge is what Christianity sanctions, the test of knowledge will be biblical citation. Each of these types of knowledge sets out a different necessary condition for knowledge in the form of justification for a knowledge claim it accepts—if there really are all these different types of knowledge.

Now the answer to the question of how many different kinds of knowledge there are is plainly an issue in epistemology, and anyone who opts for one kind of knowledge as the goal of his or her discipline implicitly adopts some stated or unstated epistemology that explains why his or her brand of "knowledge" is knowledge. However, when competing epistemologies set out different proximate goals for science, by advocating incompatible criteria of justification, which account of knowledge is the right one?

The trouble is neither can we expect philosophy to identify the correct epistemology and thereby the appropriate proximate goals for each or all of the sciences, nor can we expect the sciences to stop, pending the decision on the correct epistemology. In this respect, alternative views on the proximate goals of science, prediction, empathetic understanding, etc., reflect alternative epistemological theories that cannot be finally adjudicated now or in any foreseeable future. Thus epistemological differences are like differences in values or tastes. And, as economists are fond of telling us, there's no accounting for tastes.

If epistemological differences really are like this, current work in the philosophy of social science is condemned to persistent and irresolvable controversy. Given the way scientific goals and theories constrain methodological rules, the absence of much agreed-upon theory in the social sciences, coupled with disagreement on its proper goals, precludes any agreement on the methodological rules on which it might proceed. It does not of course follow that social science is impossible or pointless pending the resolution of these issues about its goals. Rather it means that social scientists have had to take sides on these issues; they have had to adopt explicit or implicit philosophical positions. And it is the positions on philosophical matters they have taken, in the philosophy of mind and action, on the nature of causation, and on the nature of knowledge, that locate their instrumental goals and prescribe their tentative methods.

Thus, so long as an economist identifies some goals for economics, the economist is committed to some philosophical position or other, and these goals and the methodological rules the economist endorses commit him or her to further theories, both economic ones and philosophical ones. In this sense neoclassical economics is committed to some philosophy of science or other.

THE PROXIMATE GOALS OF ECONOMICS

It seems evident that if forced to, most neoclassical economists would endorse an empiricist account of knowledge, which makes the proximate goal of science the successful testing of its claims by experience and, more specifically, by prediction. The orthodoxy of this view is reflected in the authority which the profession has accorded Milton Friedman's "The Methodology of Positive Economics,"[4] which endorses the central if not unique role of prediction in the assessment of theory. Of course, economists may not attach much importance to their endorsement of an epistemology, not thinking it very relevant to their day-to-day work as economists. If the argument of this chapter is correct, economists are wrong to think their work utterly detached from epistemology, though they are right not to suppose it has much bearing on their day-to-day concerns. However, if and when an economist steps back and considers the degree to which the discipline attains its goals, the commitment to a particular epistemology must come into play. Perhaps therefore, economists should refrain from such global assessment. Of course we know full well that whether they do so or not, others will not deny themselves the opportunity to engage in such a broad-based inquiry about the discipline of economics.

I share the economists' commitment to predictive success as at least a necessary condition of knowledge. However, it is controversial exactly what significant implications follow from this commitment about the criteria by which one should measure scientific success. In part this is because of vexed questions among philosophers of science about what constitutes predic-

4. In his *Essays in Positive Economics* (Chicago: University of Chicago Press, 1953), pp. 3–43.

tion. Following other philosophers who have written on confirmation theory, I accept that the difference between prediction and other sorts of empirical testing, like retrospective applications of new theory to old data, may not be epistemic. Prediction and retrodiction probably both provide the same amount of justification for a scientific claim. I say probably, because whether prediction proves more than retrodiction is an unsettled question in the philosophy of induction.

There are even more fundamental philosophical problems facing any simple claim that predictive success is a necessary condition of knowledge or an end to be sought above others in science. One of them is the problem of stating exactly and unobjectionably what counts as confirmation of a prediction. But this is a philosopher's problem, not an economist's. Like other social scientists, the economist knows a confirmed prediction when he or she sees one, and of course what is crucial in science is not the sheer number of confirmed predictions a theory makes, but the proportion of right predictions to wrong ones and the precision of the predictions it makes, along with the amount of surprise generated by its predictions.

The two predictive criteria of proportion of hits to misses and precision are especially important for economics. Economics is employed to provide explicit guidance to policy, both to the decisions individuals make about the microeconomic future and to the decisions governments make about the macroeconomic future. Insofar as the economists' accepted criterion for knowledge is not just successful testing but also predictive success, their epistemology is not just empiricist; it is pragmatist as well. Pragmatism in epistemology is the doctrine that predictive power is not just necessary for knowledge but is sufficient for it—or at least, what prediction cannot provide is without scientific interest. Those economists who eschew all interests beyond models that can be applied to policy are committed to such a pragmatist epistemology.

For purposes of this book, I stipulate the following implication of the economists' commitment to an empiricist epistemology: a scientific discipline should be expected to show a long-term pattern of improvements in the proportion of correct predictions and their precision. If a discipline does not show such a pattern over a long enough period, its cognitive character becomes increasingly puzzling.

Given the argument of the last section, it would be fatuous for either a philosopher or an economist to expect to substantiate a shared criterion of knowledge or a proximate goal for science. We cannot do so to the satisfaction of those with a different criterion any more than they can convince us of theirs. Epistemology is not a discipline in which consensus should soon be expected. So, having identified an epistemology widely held among economists that I share, I shall say only three things in defense of it, none of them very compelling to anyone already committed to an incompatible epistemology.

First, I recognize that there is a strong anti-empiricist tradition in the philosophy of social science and some of the social sciences themselves. Many philosophers and scientists, natural and social, identify different nonempirical proximate goals for their disciplines. Thus, for example, critical social theorists will accord prediction pride of place as a mark of knowledge in the natural sciences but require "understanding" or "intelligibility" as the mark of knowledge in their own disciplines. And some natural scientists pessimistic about the scientific credentials of the social sciences may even agree that the latter cannot aim for the sort of knowledge physics or chemistry provides— predictively useful information. The idea that what we seek in social science is knowledge certified on criteria of intelligibility or understanding is a widely shared one.[5]

But, I am constrained to ask, how do we know when we have attained understanding of an action or an institution, a historical event or a process? Can we rely on the subjective feeling of relief that comes when the hitherto unintelligible action is provided with an interpretation that makes sense? Reliance on subjective feelings is not much of a basis for a knowledge claim. As Weber pointed out long ago, such feelings need at least to be supplemented by empirical evidence. I am inclined to go further and claim that without the empirical evidence, the internal feeling of understanding is no mark of knowledge at all.

5. In fact, I discuss this view at length in *The Philosophy of Social Science* (Boulder: Westview Press, 1988), chaps. 1, 3, and 7.

Second, the proximate goal of predictive success is the only one worth aiming for by a discipline that hopes to be relevant to the guidance of public and private policy. If a certain approach cannot attain this goal, it should be surrendered by anyone who seeks practical steps to ameliorate or prevent the degradation of the human condition.

Third, many of those social scientists who repudiate predictive success because they conclude (rightly or wrongly) that their own disciplines cannot attain it would prefer it if their approach did provide predictive knowledge. They have, however, tailored their tastes in epistemologies to fit the actual character of their disciplines, as they see them.

In the rest of this book, I want to explore the degree to which we need to tailor our epistemology and the actual character of economics to fit well. How good a case can we make for the prospects of economics and in particular neoclassical economic theory without repudiating the empiricist-pragmatic aim of science? If some economists and philosophers are right, this would be an easy task, for they believe that the empiricist-pragmatic criterion of knowledge entails neoclassical theory. As we shall see, matters are not so simple. If other economists are right, the task of reconciling an empiricist epistemology with neoclassical economics is impossibly difficult, and we have to find another, more attainable goal for economics than the one this epistemology dictates. Few economists will share this pessimism, even when it is disguised as indifference to a positivist philosophy of science. However, shorn of its rhetorical baggage, perhaps there is something to the suggestion that we need to rethink our assumptions about what the goals of economic theorizing are. If we take seriously the notion that science requires persistent predictive improvement, our understanding of the aims and methods of economics becomes cloudier and cloudier, as the argument of this book reveals.

In the next chapter I examine these two alternative views of the bearing of philosophy of science on economics: the view that economics requires epistemological foundations, and the view that there is no such thing as epistemology for any science that we are required to get clear first, before assessing that science for success or failure.

2

TWO DEAD ENDS IN
THE PHILOSOPHY
OF ECONOMICS

Having explained what the bearing of philosophy on economics should be, in this chapter I examine two alternative views about this relation. What is remarkable about these two views is the fact that they are polar extremes of a spectrum of approaches from the outpouring of works on the philosophy of economics which the decades of the seventies and eighties have produced. These extremes are obvious: one might hold that philosophy is utterly irrelevant to the aims, substance, and methods of economics; alternatively, one can claim that particular economic theories are fixed, or at least very highly constrained, by "philosophical" foundations or presuppositions. Surprisingly enough, these two extreme views have actual proponents.

The extreme views need to be dealt with because they have serious ramifications for economics. The first view, that there are no problems in economics of a peculiarly philosophical sort, reflects either a deeply pessimistic or an unwarrantably complacent view about the future prospects of economics to provide knowledge. The second view is less dangerous for economics, because it is after all hard for economists to take seriously. However, this second view, that economics and philosophy are closely connected, can be dangerous, because it distracts economists who dissent from current orthodoxy in economics and shifts their attention from improvements in economics to arcane disputes in epistemology and metaphysics.

There is no need to invent the extreme views I shall treat, for they have live and indeed lively exponents well known among economists and philosophers. The view that philosophy determines economic theory is defended in the collaboration of a

21

philosopher and an economist, Martin Hollis and Edward Nell, *Rational Economic Man*,[1] and the view that economics is and ought to be indifferent to the attentions of philosophy is advanced by an economist, Donald McCloskey, in *The Rhetoric of Economics*.[2] Despite their differences, these two views share certain presuppositions. Both treat only one philosophy of science as material to their claims, logical positivism, and both repudiate it. In its place, Hollis and Nell substitute another philosophy, for they hold that an economic theory requires one, whereas McCloskey believes that the irrelevance of positivism means economics can dispense with philosophy altogether. The restriction of both of these arguments to logical positivism as their sole target is parochial but probably harmless: practically all methodological views find it convenient to define themselves by their differences with positivism. Thus we need not fear the loss of much generality by this restriction in their presuppositions.

By expounding and showing what the matter with these two views is, I hope both to reinforce the empiricist epistemological presupposition which most economists share and to reveal the seriousness of the problems which this shared presupposition makes for economics.

DOES PHILOSOPHY CONSTRAIN ECONOMICS?

Along with my *Microeconomic Laws*,[3] Hollis and Nell's *Rational Economic Man* was one of the earliest of the current generation of essays in the philosophy of economics. *Microeconomic Laws* took a relatively moderate empiricist view of neoclassical economic theory, one that had the effect of at least implicitly defending its conceptual credentials as an empirical science. *Rational Economic Man* took an extremely different view: Hollis and Nell held that "economic theories are to be judged partly by whether they are backed by a sound theory of knowledge" (p. 13) and "that neo-

1. (Cambridge: Cambridge University Press, 1975). Unless otherwise noted, all page references in the next section of this chapter are to this work.
2. (Madison: University of Wisconsin Press, 1986).
3. (Pittsburgh: University of Pittsburgh Press, 1976).

classical theories are unsound and that they rely for defense on a Positivist theory of knowledge which is also unsound. Having sought vainly for a trustier branch of empiricism, we [Hollis and Nell] shall finally argue the merits of a rationalist philosophy and a classical or Marxian economics" (p. 1).

It is more than an irony that ten years later McCloskey argued for the direct denial of this thesis about positivism and neoclassical economic theory. He claimed that the former is utterly unsuited to the latter, and therefore we can dispense with philosophy altogether in any attempt to understand what is going on in economics. It is more than an irony, because most of the reasons Hollis and Nell give for repudiating positivism, and with it bourgeois economics, are identical to the reasons McCloskey gives for preserving economic theory from severe positivist criticism as empty pseudoscience. As we shall see, McCloskey is more nearly correct about the relations between positivism and neoclassical economics. But Hollis and Nell, for all their excesses, are closer to seeing that economics, like any science, does ultimately rely, in part, on a theory of knowledge.

What can it mean to say that neoclassical economics is "wholly or partly backed" by positivism or any philosophy of science? It could mean one or more of at least three quite different things:

1. Positivist strictures on method entail neoclassical economic theory, or some important parts of it. For instance, from a distinctively positivist doctrine, like the notion that only testable propositions are meaningful, a distinctively neoclassical doctrine, like the transitivity of preferences, can be derived.

2. Neoclassical economic theory entails or presupposes a positivist philosophy of science. For instance, the impossibility of interpersonal utility comparisons might be held to presuppose an emotivist approach to ethical evaluation associated with positivism.

3. Neoclassical theory can best be defended against certain serious objections by appeal to positivist strictures. Thus, someone might defend neoclassical theory against the charge of making unrealistic assumptions by appealing to a positivist account of scientific theories that deprives these assumptions of

their literal meaning, and so blocks the objection that they are unrealistic.

Hollis and Nell do not attempt to show any of these three things. And what they in fact accomplish has quite the opposite conclusion: it shows that so far from backing neoclassical theory in any of these senses or in some further way, positivism's chief tenets are utterly incompatible with economic theory.

Now, in a way it is not surprising that Hollis and Nell do not substantiate their claim about positivism and economics under any of these three interpretations. For it just can't be done! No one can extract distinctive economic claims from positivism, or vice versa; nor does positivism provide a good, let alone the best, defense of economic theory against obvious potential objections. A brief glance at the intellectual history of the twentieth century should make this pretty clear. To begin with, economics changed less over the period of positivism's ascendancy than any other social science; economics is the discipline on which positivism has had the least effect. The characteristic feature of positivism, over the approximately forty years of its life, was its commitment to varying criteria of cognitive significance. These principles, according to which only what was testable was cognitively meaningful, had a far-reaching effect on all the social and behavioral sciences. In psychology they provided the principal weapon in the hands of behaviorists and experimentalists. They had a deciding influence on the present state of psychology as a laboratory-animal-oriented science, long opposed to any sort of systematic theorizing whatever. Similarly, in politics, sociology, and social psychology, positivist demands on testability led to new methods, like survey research, and new findings about political and social behavior, as well as new theories to explain them. But there was one great exception to the remarkable effect positivism had on the social sciences in the period of its ascendancy: economics. The style, the problems, the methods, and the concepts of economic theory have not undergone changes anything like those that overtook the other disciplines. The marginalism of Jevons, the general equilibrium of Walras, improved, enriched, and embroidered, were economic orthodoxy before, during, and after the positivist onslaught. Positivism deprived economics of none of its prejudices and certainly added nothing new by way of research methods or substantive findings. With the doubtful exception of revealed preference theory, positivism left economics just as it found it. If positivism entails or is entailed by economic theory, the economists certainly didn't notice.

It is undeniable that during the heyday of positivism, some economists (including famous ones like Friedman, Samuelson, Machlup, and Koopmans)[4] took up a garbled version of positivism, with the proportions and emphases mostly wrong, and attempted to clothe their discipline in its respectability. With the decline of positivism, these clothes, like the emperor's, disappeared. This should be no surprise, considering how ill-fitting positivist doctrine really is for economics. I will not rehearse here the conflicts between positivist prescription and economic practice. On this point chapter 7 of *Microeconomic Laws* remains unchallenged, even if the rest of the book is open to argument. In fact, the conflict between positivist prescription and economic practice was noticed by T. W. Hutchison during positivism's formative decade in *The Significance and Basic Postulates of Economics*,[5] almost forty years before the appearance of either *Microeconomic Laws* or *Rational Economic Man*. So far from supposing that positivism backed economic theory, Hutchison noted quite the opposite: positivism shows that "being unconditionally true and neither confirmable nor contradictable by any empirical synthetic proposition, propositions of pure [economic] theory cannot tell us anything new . . . about the world." They merely "call our attention to implications of our definitions."[6]

But the most remarkable feature of Hollis and Nell's argument is that, if anything, it shows that positivism *undercuts* the

4. Milton Friedman, *Essays in Positive Economics* (Chicago: University of Chicago Press, 1953); Paul Samuelson, "Problems of Methodology—Discussion," *American Economic Review* 53 (1963): 232–36; idem, "Theory and Realism—Reply," *American Economic Review* 54 (1964): 736–40; F. Machlup, *Methodology of Economics and Other Essays* (New York: Academic Press, 1978); T. Koopmans, *Three Essays on the State of Economic Science* (New York: McGraw-Hill, 1956).

5. (1938; reprint, New York: A. M. Kelley, 1960).

6. Ibid., p. 34.

claims of neoclassical theory to cognitive legitimacy! They begin
with the revelation that there is strictly no such thing as testing a
theory by comparing its predictions with the data, for strict falsification is impossible: testing always requires auxiliary assumptions to establish the initial or boundary conditions; a
disconfirmation discredits the conjunction of the hypothesis under test and these auxiliary hypotheses but does not indicate
which is at fault. This difficulty with falsification was early em-
phasized by Pierre Duhem[7] and is the cornerstone of many refutations of positivism. It undermines continuing demands for
falsifiability in economics by followers of Karl Popper.[8] Hollis
and Nell apply the lesson to neoclassical economic theory thus:

> To test a simple static proposition we would wish to com-
> pare static positions. Static positions occur in time and
> hence there must be dynamic adjustment; actual variables
> are dynamic and hence our observations are unsuited in
> themselves to test the static theory. So we must adjust our
> observations. But on what principles?
>
> Economists . . . might suggest an appeal to the "Corre-
> spondence Principle" . . . which states that the dynamic
> stability conditions of a model will determine the reaction
> of a static system to the displacement of one or more of its
> parameters. But such an appeal *must be rejected* by the high
> court of Positivism. For we can hardly assume the truth of
> the Correspondence Principle a priori and we should need
> to know what the comparative statics results were before
> we could claim to know what the dynamic stability condi-
> tions of the model were. We would have to know the equi-
> librium position before we could know anything about its
> stability. (Pp. 31–32, emphasis added)

Testing a claim in comparative statics involves taking sides on a
principle about dynamic stability. This claim too needs confir-

7. *Aim and Structure of Physical Theory*, translated by P. P. Weiner (Princeton: Princeton University Press, 1954). Originally published in 1904.

8. For a good example of such demands, advanced even in the recognition
of this problem, see Mark Blaug, *The Methodology of Economics* (Cambridge:
Cambridge University Press, 1980).

mation, and if it is false, then the apparent disconfirmation of our static claim may be attributed to its falsity. But this is an argument against the compatibility of positivism with neoclassical theory and method, not an argument for the entailment of one by the other. Its conclusion, that testing of the sort positivism demands is impossible in economics, can hardly show that there is an entailment relation between neoclassical economics and positivism. Yet this is what Hollis and Nell require.

Like others before them Hollis and Nell conclude that the positivist account of testing makes testing impossible. Unlike others, instead of casting around for an improved theory of testing, they conclude that there is no such thing as testing a hypothesis at all: "To test a theory, we must have a way of discounting the outside influence. Since this means being able to assess the influence of the exogenous on the endogenous variables, we need in effect an interdisciplinary theory, which will be more sophisticated than the theory we are trying to test and which moreover will need to be already well confirmed" (Pp. 27–28). This infinite regress condemns positivism and with it, Hollis and Nell insist, the economic theory they reject. But their entire attack on economic theory consists in showing that it fails to satisfy positivist strictures on testability. Thus, they argue that the economic agent as a rational calculator of preferences is without predictive value: the "embarrassment of [untestability] was compounded . . . by the addition of rationality assumptions. Neoclassical economics is the study of Rational Economic Man, who is not an actual man, but rather any actual man who conforms to the model. So instead of providing the missing occasion when tests are decisive, he reinforces the previous circularity" (p. 112).

The moral of the story should be either that (a) positivism is discredited as an account of the actual character of science—a conclusion already in general circulation—or (b) neoclassical economics is discredited as failing to satisfy positivist strictures. But Hollis and Nell embrace neither of these conclusions. Instead they write: "Positive economics is absurd without a Positivist justification for its methodology. Positivism underpins a gamut of opinions from the marginalist theory of the household and firm to the stock critique of the labor theory of value.

If as we are claiming, that justification is absurd, then a different philosophy is called for" (p. 42).

Hollis and Nell offer a "different" philosophy: rationalism, with its attendant a priori knowledge, and attempt to parlay this commitment into nothing less than a Marxian-Ricardian economic theory, in which production instead of scarcity is the driving force in economic behavior. But instead of pursuing this argument, we need to go back to the original claim, that neoclassical theory is backed by positivism, or any other particular philosophical foundation. That an economic theory has some philosophical foundation is something Hollis and Nell never substantiate. But as we saw in chapter 1, the assumption is correct, albeit in a weaker sense than they recognize.

Hollis and Nell identify ten "tenets" as providing the real foundations of the economic theory they repudiate. (It will be useful to compare this list with ten positivist strictures for economics that McCloskey identifies in his argument for the opposite conclusion from Hollis and Nell's.)

1. Claims to knowledge of the world can be justified only by experience;
2. Whatever is known by experience could have been otherwise;
3. All cognitively meaningful statements are either analytic or synthetic but not both;
4. Synthetic statements, being refutable, cannot be known a priori;
5. Analytic statements have no factual content;
6. Analytic truths are true by convention;
7. A known causal law is a well enough confirmed empirical hypothesis;
8. The test of a theory is the success of its predictions;
9. Judgments of value have no place in science;
10. Sciences are distinguished by their subject matter and not by their methodology. (P. 10)

Of these the first two are identified as distinctive of empiricism, the next four as characteristic of logical positivism, and the final four as implications of positivism for scientific method. Many writers who decline the label positivist willingly embrace all of these tenets. Positivists themselves would be troubled by the absence from this list of distinctive claims like the criterion of cognitive significance, the logistic character of mathematics, or the emotivist theory of ethics.

Hollis and Nell are surely correct to identify these principles as having some foundational role to play with regard to economic methodology. All of them are true, though some require qualification, others are vacuously true, and still others are in need of considerable further explication. The importance of these principles lies in the fact that they stand behind every scientific discipline. They provide the necessary conditions for knowledge in all the disciplines. As such they can hardly "back" one discipline, still less a theory in that discipline, more fully than any other. Scientific theories are accepted, rejected, modified, and expanded continually in all the scientific disciplines, without any effect on the acceptability of these tenets. It follows therefore both that they have little to tell us about the day-to-day methods of science and that these methods do not very directly underwrite or sustain the broad principles either.

Positivism represented an attempt both to increase the precision and strength of these ten principles and to derive substantive methodological dicta from them, rules that might set real restrictions on scientific method—no appeal to introspection, extensionality in concept formation, quantitative degrees of confirmation, finite axiomatizability of theories, deductive-nomological explanations, theory-neutral observation language, etc. Positivism failed to provide such principles. But its failure did not show that the general tenets, like 1–10, are wrong.

Any attempt to understand the aim and methods of science requires something like these general tenets. If we are to identify some theories as constituting knowledge, and some methods as reliable in the production of knowledge, we need at least a definition of knowledge and a generic account of how it is acquired, expressed, and certified. Certifying this generic account is itself an important philosophical problem. For our treatment of economics it is enough to note that these ten tenets (suitably amplified and qualified) are in fact the only such generic theory of knowledge that is not obviously incompatible with the actual practice of science from high-energy physics to cognitive psychology.

Whether these principles can actually help us shed light on the nature of economic theory is one of the topics of this book. But that they are either certified by or in turn certify the distinctive theses of that theory is clearly wrong. No issue in either philosophy or economics can be decided by considerations from the other discipline *alone*.

ECONOMICS WITHOUT EPISTEMOLOGY

In fact, hardly anyone who nowadays doubts the ten tenets mentioned above actually endorses their direct denials, still less their rationalist contraries. Instead, the complaint is "A plague on both your houses": science neither has nor requires any extrascientific philosophy to back it up, whether a relatively innocuous one like these principles or a more substantive one like positivism or latter-day rationalism. The fullest expression of this view among economists is to be fund in Donald McCloskey's *The Rhetoric of Economics*.[9]

I examine McCloskey's view at some length, because it is in many ways more significant than Hollis and Nell's. It is in two respects easier for economists to take seriously: first, McCloskey endorses neoclassical economics, instead of attacking it; second, his approach obviates any concern whatever with the philosophy and methodology of economics. Accordingly McCloskey's view is far likelier to win acceptance among economists eager to avoid the distractions of philosophy. *The Rhetoric of Economics* is also "trendier" than Hollis and Nell's book. Like their book it repudiates positivism, but it does so from a far more fashionable position than that of eighteenth-century rationalism. McCloskey's attack on philosophy rides the coattails of "deconstructionism." *The Rhetoric of Economics* is important, but not because of its fashionability. Deconstruction—like its predecessors "phenomenology," "hermeneutics," "structuralism," "semiotics," "Ideologie-Kritik"—will surely be succeeded by some new

9. Page references in this section are to this work.

"ism." McCloskey's view is important because it is a Sophistic invitation to complacency about economics and an attempted seduction of the discipline into irrelevancy. What McCloskey substitutes for philosophy is nothing less than abolition of economics as an organized body of knowledge. For all their errors, Hollis and Nell's views have far less detrimental consequences for economics than McCloskey's and are far closer to the truth about its relation to philosophy.

If McCloskey is right, there are no serious problems in the philosophy of economics that might infect economics itself; indeed, if he is right, the philosophy of economics does not exist. What is worse, if McCloskey's doctrine is right, there is no hope for improvement in economic knowledge—for in his view, at its best economics is already as good as we can expect it to get and should not be expected to meet the needs of practical policymakers that economists have traditionally aimed at fulfilling.

Thus, McCloskey's approach to economic methodology should be far more worrying to neoclassical economists than Hollis and Nell's attack on the credentials of their theory. If he is right, economists will have to stop doing economics or else consign their discipline to the status of a genre, a stylistic tradition in literature, like the heroic couplet or the stream-of-consciousness novel.

As will become evident, it is in the nature of McCloskey's strategy, and that of the movement throughout the social sciences that his work represents in economics, that it cannot be "refuted." The rhetorical approach proceeds on the assumption that there are no such things as good and bad argument, only persuasive and unpersuasive discourse. That is why it is a species of Sophistry. The most one can do to undermine this doctrine in a way that does not beg the question against it is to assemble a list of its most absurd consequences and invite the impartial observer to decide whether they can be tolerated. The irony is that in not begging the question against McCloskey we do him a justice he cannot even consistently identify: that of employing only valid arguments, instead of merely persuasive ones.

How does McCloskey arrive at this complacency about his discipline? Like Hollis and Nell, McCloskey recognizes that neoclassical economics and logical positivism are ill-suited to each other. Like them, he identifies a decalogue of positivist commandments which he claims economics cannot and should not satisfy. It is worth noting that his list is slightly different from theirs, and even incompatible with it in places; moreover, it is really no more representative of distinctively positivist claims than theirs is. But in one respect both lists agree. Hollis and Nell attribute to positivism the doctrine that "the test of a theory is the success of its predictions," and McCloskey heads his list of positivist commandments thus: "Prediction and control is [sic] the point of science" (p. 7). But as McCloskey notes in the first paragraph of his book, the predictive success of their science is not something on which economists agree: "Economists agree on more than is commonly understood. Their disagreements about prediction and politics give them an unhappy reputation, yet they agree on many things" (p. 3).

Hollis and Nell's response to the failure of economic theory to provide agreed-upon theory-testing predictions was to jettison both positivism and neoclassical economics. McCloskey's strategy is apparently more reasonable. In common with sociologists, anthropologists, social psychologists, and those political scientists who do not self-consciously model their discipline on economics, McCloskey puts the blame for apparent problems in his discipline on philosophy. The problems are apparent, but the blame is uncalled for, because positivism is wrong. In fact positivism is just the tip of the iceberg. It is just a symptom of a serious intellectual disorder that has held learning in its grip since the seventeenth century. The name of this disorder is modernism, "the program of Descartes, to build knowledge on a foundation of radical doubt" (p. 5).

Modernism's methodological vision, positivism, "argues that knowledge is to be modelled on the early twentieth century's understanding of certain pieces of nineteenth and especially seventeenth century physics." McCloskey's positivist "decalogue" is more familiar to social scientists than Hollis and Nell's:

1. Prediction is the point of science, only observational implications matter to a theory's truth (*sic*. This is probably a slip. It would be more accurate to substitute "warrant" or "justification" for "truth").

2. Observation means reproducible experiments, not the interpretation of people's statements about their actions.

3. Theories can be falsified by experimental observations if (and only if) they can be falsified at all.

4. Introspection is not observation, because its results cannot be independently confirmed.

5. Kelvin's dictum: "When you cannot express it in numbers, your knowledge is of a meager and unsatisfactory kind."[10]

6. The causal origin of a belief is irrelevant to its justification.

7. There is a real difference between science and nonscience, and between description and valuation, which methodology should identify.

8. Explanation is subsumption under laws and theories.

9. Scientific findings cannot justify fundamental normative claims.

Although McCloskey holds that these principles are the heritage of Descartes, he claims that economics embraced positivism only as recently as the period during which the discipline was mathematized, largely between the two world wars. Unlike Hollis and Nell, he claims that economics adopted positivism as a convenient ideology, not as a conceptual foundation or inevitable implication of economic doctrine. In this he is nearer the mark than they are. Positivism, he claims, was an ideology convenient for rationalizing the conversion of the subject from a literary to a quantitative form of presentation. But, insists McCloskey, positivism now makes economists "prone to fanaticism and intolerance."

Whether economists are fanatical or intolerant, the fault cannot be laid at the door of positivism. The doctrines of positivism were brought to America by socialists escaping Nazism and democrats escaping Soviet totalitarianism and were designed to combat fanaticism and intolerance. It was the positivists' chief aim to do so, by substituting experience for oracular authority in the certification of knowledge and by undercutting the claims of moral certainty that breed intolerance of moral differences.

McCloskey recognizes his ad hominem attack on positivism as historically inaccurate: "[A]dherence to logical positivism and

10. *Popular Lectures and Addresses* (London, 1888), p. 73.

its coreligions [might have been justified] in fascist countries between the wars and in certain workers' democratic republics
. . . . But it [is not justified] . . . in an open, plural, and pragmatic society" (p. 41). "Something is awry with [positivism's appeal for an open intellectual society], an appeal defending itself
on liberal grounds, that begins by demarcating certain means of
reasoning as forbidden and certain fields of study as meaningless" (p. 23). The defender of positivism should reply, "Not at
all," when the means of reasoning, like arguments from the authority of an infallible leader, are fallacious, and when certain
fields, like Marxism-Leninism or racist "science," turn out to
be simultaneously cognitively empty and morally dangerous.
Positivism, like other empiricist philosophies, is not a total-
itarian dictatorship. It is a government of laws, however. Such a
government alone can protect us from intellectual demagogu-
ery and "democratic centralism" in the pursuit of knowledge. If
its laws turn out to be too restrictive to permit us to attain our
scientific ends, then we need to amend them or substitute new
laws. McCloskey's solution is, to use his own description, "anarchy" (p. 40).

The actual connection between positivism and economics is
far more tangential and complex than McCloskey suggests.[11]
And it has only a little to do with the mathematization of the discipline. As noted above, positivism's first real brush with economics occurred in the pages of T. W. Hutchison's attack on the
scientific credentials of the subject, in *The Significance and Basic
Postulates of Economics*. The enthusiasm of economists since
Walras for mathematics certainly does stem in part from economists' admiration for the predictive powers of nineteenth-
century physics and the belief that these powers are conveyed to
it by the expression of its leading ideas in differential equations.[12] Equally important to the motivation to mathematize the

11. Probably the fullest and most philosophically sophisticated account of
this connection is to be found in chapter 8 of Daniel Hausman, *The Separate
and Inexact Science of Economics* (Cambridge: Cambridge University Press,
1991).

12. For a well-informed though highly polemical account of the influence
of nineteenth-century physics on twentieth-century economics see Philip Morowski, *More Heat than Light* (Cambridge: Cambridge University Press, 1989).

subject was Walras's achievement of transforming Adam Smith's claims about individuals and markets into something approaching an axiomatic system and its theorems.

In fact a compelling case can be made for the claim that the alliance between positivism and "positive economics" stems from the economists' recognition that this doctrine explained and justified their self-imposed demand that economic knowledge have the kind of predictive content that would suit it to the guidance of policy. In Hutchison's work, and in others, positivist methodology was in fact employed to downgrade the importance of mathematics in economics, not to rationalize it. Hutchison's attack on "pure theory," as he called it, was an attack on mathematical economics and the deductive reasoning through which it proceeds. More sophisticated appreciations of positivist doctrine by subsequent economists writing on methodology showed that these criticisms were misguided, and that mathematizing economics was consistent with positivism's demands on theory. But they never drew the fallacious conclusion that mathematizing economics was required by positivist strictures in philosophy.

If there is any ulterior "ideological" motivation in economists' embrace of positivism it was probably *not* the notion that the results in the philosophy of science endorsed neoclassical economics, but rather the conviction that the results stigmatized Marxian economics, whether mathematical or literary, as pseudoscience, and institutional economics as mere chronicle or taxonomy. Marxian economics could be consigned to the wastebin of science because its claims about the labor theory of value were untestable and its account of social chance was "historicist" and ipso facto wrong. Institutional economics sought no general laws to explain its claims, and so could not be called science. Thus positivism rationalized neoclassical economics to the extent that it undermined the claims of its competitors.

But what exactly is amiss with what McCloskey calls modernism and its methodological offspring? What renders it unsuitable as an account of either what economists actually do or what they should do? Like Hollis and Nell, McCloskey identifies two defects in modernism: the impossibility of falsification and the superfluity of prediction as a scientific aim. His objection to falsification recapitulates the arguments of Duhem noted above. McCloskey's second claim is far more serious: many important scientific theories make no predictions; therefore, the absence of well-confirmed predictions in economics is no sign of scientific inadequacy. Positivism is wrong to set out prediction as a goal of science. If it were right, claims McCloskey, we should have to stigmatize the work of no less than Darwin as unscientific:

> It is a cliche among philosophers and historians of science . . . that one of the most successful of all scientific theories, the theory of natural selection, makes no predictions and is therefore unfalsifiable by prediction. With fruit flies and bacteria, to be sure, one can test the theory in the approved manner; but its main facts, its dinosaur bones and multicolored birds, are things to be explained, not predicted. Geology and evolution, or for that matter an astronomy of objects many light years away, are historical rather than predictive sciences. (P. 15).

This claim, that prediction is both unnecessary and absent in many theories, is central to the whole edifice of McCloskey's postmodernism. It is also one of the chief tenets Hollis and Nell identify in the positivist methodology they wish to reject. Here again, despite their very great differences, these polar extremes share important assumptions in common. But the repudiation of prediction is completely wrong, both in its facts and in its inferences, as I shall show in the next section. Without contesting it for the moment, let us consider what McCloskey makes of it.

Not only are many important scientific theories and disciplines blissfully innocent of predictive claims, but according to McCloskey the demand that they provide predictions would be their death knell.

> A modernist methodology consistently applied . . . would probably stop advances in economics. Ask any economist. What empirical anomaly in the traditional tale inspired the new economic history of the early 1960s or the new labor economics of the early 1970s? None: it was merely the realization that the logic of economics had not exhausted itself

at conventional borders. What observable implications justify the big investment of economic intellect since 1950 in mathematical general equilibrium theory? For all the modernist talk common among its theorists, none; but so what? Could applications of economics to legal questions in the style of the emergent field of law and economics rely entirely on objective evidence? No; but why would one wish to so limit the understanding? . . . There is nothing to be gained and much to be lost by adopting modernism in economics. (Pp. 18–19)

Instead of modernism McCloskey defends a methodology, or rather an antimethodology, that he sometimes calls "pragmatism" or "anarchism" (p. 29) but mainly refers to as "the rhetorical approach." The irony of calling a philosophy that downgrades predictive success "pragmatism" seems lost on McCloskey. "Rhetoric does not deal with Truth directly; it deals with conversation. It is, crudely put, a literary way of examining conversation, the conversation of economists and mathematicians as much as of poets and novelists. It can be used for a literary criticism of science. The humanistic tradition in Western civilization, in other words, is to be used to understand the scientific tradition" (pp. 28–29).

Since McCloskey says that rhetoric does not deal with truth directly, the natural implication is that it deals with truth indirectly, and so holds out the prospect of improving economics, not just rationalizing its status quo. But it is clear from McCloskey's own account of rhetoric that it cannot hope to help foster any improvement in the cognitive merits of economics. It can only help improve the marketing skills and public relations of economists with one another and with noneconomists. "Rhetoric is the art of speaking. More broadly it is the study of how people persuade" (p. 29). But this is nothing more or less than the art of the Sophists, like Protagoras, who made men, instead of logic and the facts, the measure of what is true and what is false. It is a doctrine according to which the product with the most effective advertising campaign is the best purchase.

As if afraid of the real meaning of the rhetorical approach, McCloskey goes on from his definition of rhetoric as the study of how people persuade to quote other definitions of the discipline, ones which protect it from the charge of Sophistry, by converting it into good old-fashioned modernist epistemology. He quotes the well-known literary critic Wayne Booth's account of rhetoric:

> [It is] "the art of probing what men believe they ought to believe, rather than proving what is true according to abstract methods"; it is "the art of discovering good reasons, finding what really warrants assent because any reasonable person ought to be persuaded"; it is "careful weighing of more-or-less good reasons to arrive at more-or-less probable or plausible conclusions—none too secure but better than would have been arrived at by chance or unthinking impulse"; it is the "art of discovering warrantable beliefs and improving those beliefs in shared discourse." (P. 29)[13]

But this passage, with its emphasis on justification instead of impulsion, good reasons as opposed to psychologically effective ones, what reasonable people ought to believe instead of what fallible people do believe, is just what McCloskey rejects as modernism. He cannot help himself to Booth's definitions if he wants a doctrine detectably different from the view he sets out to attack. And he has such a doctrine, for the details of the rhetorical approach as McCloskey lays them out have noting to do with the "art" Booth identifies in the quoted passage.

To see this, consider for a moment McCloskey's examination of the rhetoric of Robert Solow's "Technical Change and the Aggregate Production Function."[14] McCloskey shows that this influential paper employs the four "master tropes" recognized in literary theory: metaphor, metonymy, synecdoche, irony. When McCloskey is finished with his literary analysis, the modernist is inclined to yawn and say, so what? Are these papers well argued as well as persuading. Do they actually justify their conclusions as well as expound them by artful means? Do they fol-

---

13. Quoting from Wayne Booth, *Modern Dogma and the Rhetoric of Assent* (Chicago: University of Chicago Press, 1974), pp. xiii, 59.
14. *Review of Economics and Statistics* 39 (1957): 312–20.

low principles of reasoning that avoid fallacies as well as conventions of style that foster conviction?

But the modernist's yawn and his rhetorical questions would be brought up short. For they betray a fundamental misunderstanding of McCloskey's point. There isn't anything more to science than effectiveness in persuasion, artful exposition, elegant fulfillment of artistic conventions. To search for more is to betray a facile misunderstanding, fostered by outmoded modernism: "Economics . . . isn't science. . . . But then neither are the other sciences. Economists can relax. Other sciences, even the other mathematical sciences, are rhetorical." Or again, "Economics is a collection of literary forms, not a science. Indeed, science is a collection of literary forms, not a science. And literary forms are scientific" (p. 55). For 'science' means nothing more or less than 'disciplined inquiry'. "The claim here is not the vulgar figure of logic that economics is mere humanism because it is a failure as a science. The claim is that all science is humanism (and no 'mere' about it) because that is all there is for humans" (p. 57).

It would be a fallacy, if there are fallacies, to infer from the fact that good science is persuasive discourse the conclusion that persuasive discourse is good science, that all there is to science is the production of artful and affecting thoughts. That may be all there is to literature, but there is more to science.

McCloskey denies this. He rejects the claim that we should "have a standard of Truth beyond mere rhetoric . . . [that] we should aspire to more than mere persuasion" (p. 42). There isn't anything more than "what persuades well-educated participants in the conversations of our . . . field" (p. 46). "There are particular arguments good and bad. After making them there is no point in asking a last summarizing question: 'Well, is it True?' It is whatever it is—persuasive, interesting, useful and so forth. . . . Truth is a fifth wheel" and persuasion is social (pp. 46–47). McCloskey's failure here is not to realize that the question of whether a claim is true or not goes unasked in science, only because there is a connection between good argument and truth, an inductive connection, but a real one for all that. If there were none, then McCloskey would be right: there would be nothing to argument but persuasion; in particular there would be no such thing as a good argument or a bad one, just a convincing argument or an unconvincing one. At any rate this is what a modernist believes. One of the special virtues of McCloskey's doctrine is that there is no such thing as a sound argument from the modernist's viewpoint, only one that may or may not convince the postmodernist.

McCloskey thinks he can persuade us that what makes science good is the artfulness of its presentation, instead of the warrant of its argument. He demonstrates that economists generally employ rhetorical devices as well as considerations that modernists would accept as evidence, and he "argues" that the rhetorical devices are more effective. But, if there are standards of reasoning beyond merely securing assent, then at most this would show that economists are fallible thinkers. Like others, economists allow themselves to be swayed by factors that lack probative force—unless of course it turns out that the reason rhetorical devices have persuasive force is because they also have probative merit. But probative merit is not an explanation for the persuasive power of the economists' traditional dialectical ploys that McCloskey can in all consistency accept.

The same problems vitiate McCloskey's conclusions about the status of the law of demand. McCloskey purports to identify eleven different good reasons economists believe that demand curves slope downward. Of these only one is alleged to be respectable by modernist standards, and the other ten are artistic and literary. They include introspection, common sense, arguments from authority, "the lore of the academy," "mere definition," and literary analogy. "Does this leave economists uncertain about the law of demand?" he asks rhetorically. "Certainly not. Belief in the law of demand is the distinguishing mark of the economist, demarcating him from other social scientists more even than his other peculiar beliefs. Economists believe it ardently. Only some part of their ardor, therefore, is properly scientific" (p. 59). What is this supposed to show? It cannot be that these ten sorts of reasons are really good reasons (though any dispassionate reader would recognize most of them as such), for McCloskey has abdicated the right to identify some reasons as good and others as bad. This argument could constitute a defense of economists' commitment to the law of demand against the charge of overreaching the evidence only on the condition that whatever actually works as a persuasive tool in economics is a good reason. Such a condition, of course, is Protagoras's Sophistry.

The eleven reasons McCloskey offers for why economists believe the law of downward-sloping demand are in sum pretty weak justification for this belief. The weakness of these grounds should not really be surprising, since the law of demand, for all its centrality to economic theory, is really not part of a very well established theory with great explanatory and predictive power. And if economists believe the law of downward-sloping demand as fervently as McCloskey suggests, then the strength of their belief is as much of a puzzle about economics as it is a difficulty for a philosophy of science which suggests that the law remains insufficiently established to warrant our credence.

Having identified reasons why economists in fact *do* believe the law of demand, instead of reasons why they *should* do so, McCloskey goes on to identify the rhetorically effective features of a number of classical pieces of economic theorizing. His examples include work by figures like Samuelson, Becker, and Muth. What this examination shows is in fact how little by way of justification the actually effective means of persuasion provide. McCloskey chooses two pages drawn at random from Samuelson's *Foundations of Economic Analysis*[15] and invites us to "consider how he actually achieves persuasion": (1) using mathematics, which, though "pointless" on the pages in question, is "warrant of expertise"; (2) making six appeals to authority; (3) employing "mere speculation" about the effects of relaxing certain theoretical assumptions; (4) making "several appeals to hypothetical toy economies"; and (5) using terms like "friction" metaphorically (pp. 70–72).

This analysis should certainly not increase our confidence in the warrant of economic theory, yet McCloskey points to the methods with some pride. For they substantiate his claim that economists, like other writers, attempt to persuade their readers. But no one has ever denied this. What has been denied, since

15. (Cambridge: MIT Press, 1947).

the time of Socrates, is that persuasion is ipso facto justification. Though he never openly admits that there is a difference between persuasion and justification, in the end McCloskey may not be quite enough of a postmodernist. He concludes his book by answering a question he poses for himself: "Suppose that modernism is dead, that economics is rhetorical. . . . So what?" His answer presupposes that there is some nonrhetorical measure or standard against which economics is to be measured, and that against that measure economics "is pretty well off." Yet he does not recognize this answer is a piece of modernism: "One thing is clear; the absorption of rhetorical thinking in economics will not precipitate any revolution in the substance of economics. Rhetoric does not claim to provide a new methodology, and therefore does not provide formulas for scientific advance. It believes that science advances by healthy conversation, not adherence to a methodology" (p. 174).

Notice the explicit commitment to scientific advance, not just to scientific change. What does advance come to? Is it correspondence with economic reality, improved predictive power? It cannot be either of these, for the very application of the words 'truth' or 'prediction' to a discipline is inconsistent with a thoroughgoing postmodernism. To advance is to improve, and to improve is more nearly to approach some goal, standard, or measure of success. In literature, there is no such thing as advance in this sense: no one has advanced on Homer despite the three-thousand-years-long "conversation" he initiated. But if there is advance in economics, it cannot be mere consensus among the persuaded, can it?

According to McCloskey, "economics at present is, in fact, moderately well off. It may be sleepwalking in its rhetoric, but it seems to know in any case approximately where to step" (p. 174). That is, economics is moving in the right direction. But this means there is a right, and a wrong, direction. How can McCloskey tell, unless there is a relation between persuasion and justification? And what is the right direction anyway? The one McCloskey favors? Can he give us an argument for his preferred direction, and will he assert that the argument is not mere rhetoric but justifies his conclusion as more likely to be true than other claims about economics?

McCloskey concludes that "the main achievement of economics is not the prediction and control assigned to it by modernist social engineering, but the making sense of economic experience" (pp. 174–75). Here at last we have the most basic contrast in modernism's philosophy of social science: that between prediction and understanding. Philosophers and social scientists have long struggled over the question of whether the social sciences must choose between affording interpretative understanding of human action—making sense of it—or subsuming it under predictive laws. Neither side in this debate ever surrendered the aim of providing knowledge—that is, justified true belief—about human action; both recognized that fallible people can be persuaded by artful though unjustified theory. In taking the side of interpretive understanding, McCloskey joins a long line of thinkers squarely in the modernist tradition beginning no later than Hegel and Dilthey. Postmodernism is just the latest version of these nonempiricist philosophers' and social scientists' attempts to undercut the epistemic goals of their empiricist opponents. To the extent that it has believed the adoption of the epistemology of the Sophists to be necessary for taking issue with empiricism, the rhetorical approach is no advance on its predecessors.

A later work, *If You're So Smart*,[16] reveals McCloskey's commitment to the separation of knowledge and prediction, along with his acceptance of economics as a genre instead of a science:

> Grown-up economics is not voodoo but poetry. Or, to take other models of maturity, it is history, not myth; politics, not invective; philosophy, not dogma. A correct economics—which is to say most of the rich conversation of economics since Adam Smith—is historical and philosophical, a virtual psychoanalysis of the economy, adjusting our desire to the reality principle. On this score Marxian and bourgeois economics can be similarly childish in giving in to temptation. A Marxian economist of an old-fashioned sort trumpeting the predictive power of Marxism makes the same childish error as does a badly educated main-

16. (Chicago: University of Chicago Press, 1990), p. 109.

stream economist thinking the future of grain prices is
predictable. A grown-up epigram would be: The point is
to know history not to change it. The best economic scientists, of whatever school, have never believed in profitable
casting of fores [i.e., forecasting].

Leaving aside for later discussion McCloskey's claim that economics cannot predict, note the implication of the last two sentences that we can have economic knowledge without subjecting
it to predictive test. One cannot change "history" without predicting what will happen as a result of one's interventions, and
what will happen if one does not intervene. If neither of these
things is necessary for establishing whether our theories constitute knowledge, or provide understanding or render the economy and its parts intelligible, then it must be because the right
epistemology can certify theories as knowledge on the basis of
other criteria—unless there is no epistemology, in which case
McCloskey has no intellectual right to claim that the point of
economics is to know anything.

Differences in epistemology thus turn out to be the crux of
the choice between modernism and postmodernism, just as it
was before empiricism and non- or anti-empiricism acquired
these new names. But if anything, McCloskey's arguments
against imposing the requirement of predictive success on disciplines that claim to provide knowledge are far more superficial
and far less convincing than the traditional ones. The reason
is that traditional arguments to this effect began by admitting
that prediction was crucial to science but insisted that the study
of human action was not science. McCloskey cannot employ
this argument, for according to his view, "science isn't science
either."

THE ROLE OF PREDICTION IN BIOLOGY, PHYSICS, AND ECONOMICS

Recall McCloskey's claims, quoted above, that important scientific theories make no predictions. His examples included geology, astronomy, and, most important of all, evolutionary
theory. McCloskey's allegation that evolutionary theory makes

no predictions is especially important. Throughout *The Rhetoric of Economics* he recurs repeatedly to comparisons between this single theory and the whole discipline of economics. Along with geology and astronomy, McCloskey calls evolutionary theory a historical, rather than a predictive, science. That evolutionary theory is not a predictive theory is practically McCloskey's sole argument for waiving prediction as a reasonable demand on scientific theories. In so arguing, McCloskey indulges in a fiction along with defenders of other predictively weak theories, like intentional psychology, for example.

These claims about evolution, geology, astronomy, and historical and predictive sciences are so riddled with confusions it is hard to know where to begin. And perhaps it would not be worthwhile but for the fact that when it comes to prediction, McCloskey's mistakes are shared by others, even those beyond the thrall of postmodernism.

First of all, pace McCloskey, it is no "cliché" among philosophers of science that the theory of evolution makes no predictions. Indeed this question of whether or not it does is one of the most controversial issues in the philosophy of biology. And for this reason the theory has been subject to more criticism—positivist, postpositivist, and antipositivist—than any other important scientific achievement. More important, most of the complaints about its lack of predictive power have either betrayed a misunderstanding of the theory of natural selection or reflect simple mistakes about the predictive content of any theory. McCloskey makes some of both of these sorts of mistakes.[17]

No theory has predictive content all by itself, for no theory contains claims about the initial conditions to which it is applied for the generation of predictions. Theories have predictive content only when conjoined to such initial or boundary conditions. The problem with the theory of natural selection is the difficulty of completely specifying the initial conditions we require in order to apply it to make predictions. Moreover, once these

---

17. The bearing of the theory of natural selection on economics is discussed at further length in chapter 6. For a full discussion of the theory of natural selection and its predictive content, see A. Rosenberg, *The Structure of Biological Science* (Cambridge: Cambridge University Press, 1985), chaps. 5–7.

conditions are specified, it is in the nature of the theory that it makes only generic, not specific, predictions. Given variation and heredity, it tells us there will be selection for a given trait, but not how much or how long it will take. For this information we must look to other theories in biology.

It is crucial to keep in mind that there are other theories in biology: theories of heredity, physiology, ecology, behavior, which systematize varying kingdoms, phyla, families, and species. When one whole discipline like economics is compared to one important but limited theory in another whole discipline, the conclusion is more than likely to be an overhasty generalization, even if the theory is correctly understood.

McCloskey concedes that we can test the theory of evolution "in the approved manner" by making predictions about fruit flies and bacteria. "[B]ut its main facts, its dinosaur bones and multicolored birds, are things to be explained, not predicted." First of all, we cannot test its claims about fruit flies and bacteria "in the approved manner" if by that he means the construction of falsifying experiments. For there are no such things. What is more, the actual tests of the theory of natural selection in the laboratory are fraught with difficulties and doubts. Is the model system we employ appropriate, is the experimental setup closed, are we allowing enough time for evolutionary effects, and most important, are we modeling anything like the actual conditions in which evolution by natural selection does take place? Despite these difficulties, we can make generic predictions and retrodictions about fruit flies and bacteria. But we can also make such predictions about larger animals and retrodictions about extinct ones, if we are given data about initial conditions. Second, McCloskey's supposition that dinosaur bones and multicolored birds constitute "the main facts . . . to be explained by evolution" is jejune at best. Who is to say what the main facts of evolution are? Except for the small minority of paleontologists, biologists care greatly about the application and confirmation of the theory of natural selection and almost nothing about dinosaur bones. It's as if a trip to the top floor of the American Museum of Natural History should determine what the crucial problems of evolutionary biology are. Like many nonbiologists, McCloskey confuses the actual course of evolution on this planet with a general theory of evolution, one which if true is true everywhere and always, here and across the galaxy. A general theory cannot account for dinosaur bones, because it is not committed to the existence of dinosaurs. And terrestrial evolution is a sequence of particular events, not a theory at all.

Leave aside McCloskey's distinction between explanation and prediction, and consider what a "historical" science could be, if geology, evolution, and astronomy are to be our examples. Either astronomy is the data gathering that astrophysics and cosmology require in order to test the generalizations of high-energy physics and relativistic mechanics, or it is the conjunction of that data gathering and the theorizing that attempts to explain the data. If it is the data alone, then its failure to predict is just a consequence of the fact that nothing follows from a description of the location of heavenly bodies at one time for their location at other times. If it is theory plus the data, then astronomy is full of predictions and retrodictions about stars, quasars, and quarks and gluons for that matter. And there is nothing historical about this science unless you mean simply that its data are dated so that we can test dynamic theories. This is neither the sense of historical science current in the philosophy of history nor the sense McCloskey requires.

As for geology, it is historical insofar as it involves the application of quite ahistorical physical principles to explain a particular sequence of events on and underneath the surface of this particular planet and to predict, for instance, earthquakes and the location of oil, as well as where the continents will be a hundred million years from now. Geology is a historical science in that geologists are interested in applying ahistorical physical laws to contemporary data in order to retrodict the sequence of events in the earth's past, and in the sense that the data we need to make geological predictions must be classified and categorized into dated kinds (Devonian, Precambrian). We need history in order to make predictions that test the ahistorical generalizations of physics which the study of geology employs. It is, moreover, a "cliché" among philosophers of science that geology is not a pure science, with its own autonomous laws. Rather, it is an applied one, because it is concerned with the course of events in a spatiotemporally restricted part of the universe. Economic history bears a parallel to geology, but only if we can parallel economic theory to physics, a decidedly ahistorical subject. But this is just the parallel McCloskey excludes as inappropriate.

What would a historical science be? The best philosophers have done in characterizing such a discipline is this: a historical subject is one in which the theoretical or explanatory principles themselves must make reference to particular events in the past. Explanations are essentially historical when a complete specification of the present state of a system is in principle insufficient to determine any future state of interest and when knowing its past states is at least necessary. In this sense, for example, Newtonian mechanics is not a historical science, because from a description of the current state of a Newtonian system and from Newton's laws, all prior and future states are deducible. It should be evident that neither evolutionary theory, geology, nor astronomy are historical sciences in this sense. Whether economics is a historical science in this sense is another matter. Certainly the microeconomic theory that McCloskey wants to preserve from the criticism of modernism is not a historical science in any sense of that term yet identified in the philosophy of science.

It is inappropriate to compare the entire discipline of economics to one particular theory in biology. At best, comparisons should proceed science by science or theory by theory. But of course McCloskey could hardly say that the entire discipline of biology makes no predictions. Yet that is the claim he requires in order to effect his comparison. Perhaps most important, a comparison between the role that evolutionary theory plays in biology as a whole and the role that, say, neoclassical microeconomic theory plays in economics will show that there are no analogies here in which economists can take much consolation. Evolutionary theory is a generic account of a mechanism—variation and selection—whose details are given by a myriad of other lower-level theories and findings in biology. But microeconomic theory is the fundamental underlying mechanism that a myriad of other findings and theories in economics require to provide their detailed mechanism. What is permissible by way of predictive weakness in a generic theory is hardly acceptable in a fundamental one. The relation between evolutionary theory and economics is explored more fully in chapter 6.

Independent of McCloskey's mistakes about biology, astronomy, and geology, there are actually three separate and general points to make about the claim that science does not require prediction. One is the relatively trivial point that practically any body of thought can be called a "science." The word has acquired both honorific status and a certain vacuity over the years: so much so that some wags have claimed that any subject with the word "science" in its title (with the possible exception of computer science) isn't one (cf. administrative science, military science, library science, mortuary science, food science, etc.). Moreover as Continental intellectual traditions become more fashionable in the English-speaking world, more and more humanists will come to style their disciplines as part of the *Geisteswissenschaften,* the "mind-sciences," as that term is best translated from the German. But as Lewis Carroll said, calling a dog's tail a leg doesn't make it one.[18]

The second point is that the best and most prized of scientific theories are in fact those with the greatest predictive power. Indeed, the history of physical science, if it shows anything, shows that scientists are prepared to sacrifice almost anything to increase the predictive power of their theories, and that they are willing to swallow all manner of intellectual discomforts in order to acquire it. Perhaps the best example of this willingness to sacrifice explanatory power, mathematical rationality, and even intelligibility on the altar of predictive power is quantum electrodynamics.

18. McCloskey thinks this terminological adoption is an argument in his favor, showing that our English-speaking distinction between science and the humanities is a parochial and a waning one. Even if it were true, it would at best be a very weak argument for the claim that some sciences need not be predictive, for the reason Lewis Carroll gave. And contrary to McCloskey's claim, the French term *"science humanes"* has long denoted, not the humanities, but the behavioral sciences. See, e.g., Gilles Gaston Granger, *Pensée humane et le science d'homme* (Paris: Aubier, 1983), translated by A. Rosenberg et al. as *Formal Thought and the Sciences of Man* (Dordrecht: Reidel, 1983); or Raymond Boudon, *A quoi sert la notion de structure?* (Paris: NRF, 1968).

This theory systematizes all physical phenomena whatever, with the exception of gravity and the forces holding together atomic nuclei. Its predictions are accurate to at least eleven decimal places. As Richard Feynman notes, "To give you a feeling for the accuracy of these numbers, it comes out to something like this: if you were to measure the distance from Los Angeles to New York to this accuracy, it would be exact to the thickness of a human hair."[19] On the other hand, the theory purchases this predictive accuracy at a high price. To begin with, it rules out as not further explainable the probabilities it assigns to the varying possible results of electron-photon interactions. As Feynman writes, "Nature permits us to calculate only probabilities." And yet despite its fantastic predictive powers, the theory cannot be correct, for unless certain arbitrary limitations are imposed on the calculations the theory requires, the theory makes neither physical nor experimental sense. Feynman calls these arbitrary mathematical maneuvers a "shell game" and he has a right to, because he received the Nobel Prize for inventing them. The technical name for the mathematics required to make the predictions come out coherent, let alone good to eleven decimal places, is "re-normalization." Feynman writes, "But no matter how clever the word, it is what I would call a dippy process! Having to resort to such hocus pocus has prevented us from proving that the theory of quantum electrodynamics is mathematically self-consistent. . . . I suspect renormalization is not mathematically legitimate" (p. 128). But, on the other hand, he insists, "the rest of physics has not been checked anywhere nearly as well as electrodynamics" (p. 131).

The moral of the story is that reports of the waning of prediction as the most important goal in science have not yet reached physics. Of course this does not show that McCloskey is wrong about prediction. After all he did not say that no science is predictive, only that lots of respectable ones aren't. But it does cast a pall over the view, common since the beginning of Kuhn's popularity, that even in physics theorizing is not controlled very largely by its predictive reliability. In contemporary physics at

19. *QED* (Princeton: Princeton University Press, 1985), p. 7. Page references in this paragraph are to this work.

any rate, such reliability seems important enough to overcome every other potential objection to the acceptance of a theory. To the extent that arguments for the unimportance of prediction in the softer sciences rely on allegations of its unimportance in physics, such arguments rest on a false premise. But McCloskey continually cites these Kuhnian claims about the "hard sciences" in order to deny any real demarcation between them and other disciplines.

There is something a bit simplistic about demanding that the summum bonum of science is everywhere and always predictive success. It is particularly vexing for philosophers of science that we do not even yet really understand what predictive success is, at least with the kind of exactness we should like. As noted in chapter 1, philosophers have as yet been unable to fully understand the concepts of testing, confirmation, and their kindred notions. There is, for example, no agreement on why the prediction of a new and surprising result affords a theory greater confirmation than the deduction of an already-known experimental result. The logical relations between hypothesis and evidence are the same in both cases, and yet many hold that new predictions are worth more than simply fitting an equation to a previously collected body of data. What seems safe to say is that what science demands by way of predictive success is at least systematic improvement of earlier predictions and at most the prediction of a startling and quite unexpected occurrence with a great deal of precision. These minimum and maximum demands operate in science today as much as they did in the seventeenth century. Among philosophers of science in recent years at any rate, the most central debate, between realism and antirealism, is about whether prediction of observations is merely necessary for knowledge, as realists hold, or all we need from science, as antirealists hold.

But besides these considerations about the role of prediction in science, the third and most important point is that although science may or may not require prediction, we do. It is a large part of what we look to science for. It is what we need to ameliorate the human condition or at any rate to prevent further deterioration in it. This applies to each of us as individuals and to the governments that regulate us. If economics cannot or need not give us predictions we can use as a guide to actual decisions, then we shall look elsewhere for such guidance. There are of course several disciplines and theories waiting in the wings, ready to supplant conventional economic theory as the basis of policy analysis and implementation. Economics cannot deny them the aim of prediction no matter how firmly it might deny this aim to itself.

Now the prospects of losing the attention of the Federal Reserve may leave some economists indifferent, and if surrendering the goal of prediction has this price, they may be willing to pay it, for then their science will no longer be distracted from its real aims by impossible demands. But I dare say, most economists will not want their field to be relegated to the status of a *Geisteswissenschaft*. However, McCloskey seems thoroughly committed to a policy-irrelevant economics.

McCloskey argues "that the literal application of modernist methodology cannot give a useful economics" (p. 16), and he goes on to catalogue examples of "advances" in economics that would never have occurred had modernism really held sway as the actual method of economics. I have quoted McCloskey's list above. It includes general equilibrium theory, law and economics, "the new labor economics of the 1970s," and economic history. I discuss some of these initiatives in chapters to follow. But initiates will recognize each of these "advances" as the formalization, extension, or application of the traditional formalisms of neoclassical theory. If these are examples of advances in economics, a critic of economics could be excused for concluding that "advance" means nothing more than old wine in new bottles.

As exponents and opponents of each of the demarches McCloskey cites will confirm, the disputes about each of them simply recapitulate the classical substantive and methodological problems of neoclassical theory. The litany of advances McCloskey cites is nothing more or less than the reiteration of the same old theory, with no thought given to its vindication at home or in its newer domains of application. The hard questions, of whether judges or slave owners or part-time workers or agents in a market approaching perfect competition and general equilibrium really do solve problems of constrained maximization, are left untouched. These demarches constitute advance only in the sense that economics has not surrendered its received wisdom in the face of its predictive weakness.

At a minimum, McCloskey owes us an account of what "useful" means and what constitutes an "advance" within the purview of the rhetorical approach to economics, or any discipline for that matter.

Quite independent of his misreading of the importance of prediction in science, McCloskey has another reason to claim that it is unnecessary in economics: it is unnecessary because it is impossible. He cites with approval von Mises's dictum that predicting the economic future is "beyond the power of mortal man."[20] The reason McCloskey gives is that if economics could predict, economists or at least those who hire their services would be rich. Any economist who thinks his or her science is predictive "must answer the American Question: if you're so smart, why ain't you rich?" (p. 16). This question becomes the leitmotiv of McCloskey's subsequent work, *If You're So Smart*. He insists that some economists allow themselves to be paid cash to answer questions like "what's going to happen to interest rates next year?" "[B]ut they know they can't. Their very science says so."[21]

To noneconomists this may sound like a joke. To economists it must sound like a ringing endorsement of the strongest version of rational expectations theory, one that claims not only that governmental economic policy is impotent but also that individual policy is as well. Concerning this view, to be examined more fully in the next chapter, economic theory tells us that economic agents have an incentive to inform themselves about both economic theory and the economic plans of government and other economic agents. These plans are based on predictions. But when all agents are informed about these plans and predictions and know the relevant theory, the reactions of at least some agents to these plans and predictions will result in the predictions being falsified and the plans going awry. Thus

20. Ludvig von Mises, *Human Action* (New Haven: Yale University Press, 1948), p. 867.

21. (Chicago: University of Chicago Press, 1990), p. 98.

economic predictions are reflexive in their effects: they undermine themselves.

The reflexiveness of economic predictions may be part of a good argument for surrendering the demand that economic theory, as it now stands, make successful predictions. It does not rest on dubious premises drawn from a misreading of the history of science or the epistemology reflected in the latest trends in literary criticism. Rather, it rests on the solid premise that 'ought' implies 'can' and that its contrapositive 'can't' implies 'need not'. If economics cannot provide improvements in prediction, then it is fatuous to demand that it do so. But independent of the lively dispute about whether economics is in fact reflexive to anything like the extent McCloskey or the most ardent rational expectations theorist supposes, this argument raises two questions:

First, isn't the very argument McCloskey makes self-refuting? After all, it issues in a prediction about the future of economics which rests on a piece of economic theorizing. McCloskey tells us that an economic argument can show that predicting the economic future is impossible: economists can't get rich, because, as "an economist would put it, in his gnomic way, at the margin (because that is where economics works) and on average (because some people are lucky), the industry of making economic predictions . . . earns merely normal [i.e., zero] returns" (p. 16). What is this but an economic prediction, deduced from economic theory, just what McCloskey tells us is impossible![22] A more defensible conclusion from McCloskey's premises is that there is a severe limitation on economic prediction: negative predictions—claims about what will not happen—are easy, as are retrodictions—fitting curves to past data; but beyond a point we approach early in economic theorizing, no improvements in positive prediction of what will happen are forthcoming. The reason is that rational agents have an incentive to employ improved predictions in a way that falsifies them after all.

Meanwhile, McCloskey faces a second, and most serious, problem: the conclusion that economic prediction is unnecessary in economics because it is impossible can hardly strengthen

22. Here I am indebted to Daniel Hausman.

McCloskey's case for its irrelevance. Its impossibility is the crucial premise in an argument that we should forgo economics owing to the fact that it cannot, even in principle, provide us with the sort of knowledge we require. No matter how many sorts of knowledge there are—*Geisteswissenschaft, Naturwissenschaft,* humanistic, scientific, interpretative, instrumental—the kind we look to economics for is the kind controlled and certified by a reality that neoclassical economics did not create, but found preexisting. Certifiable improvements in knowledge of this reality can derive only from predictions, successful and unsuccessful. And this piece of epistemological wisdom will be true no matter how convincingly and conveniently the Sophist argues that the strength of conviction, and not the way the world is, is the measure of what is true and what is not.

As I admitted in chapter 1, there is no likelihood of forthcoming settlement of the epistemological dispute about the nature of knowledge and the bearing of improvements in prediction on the certification of knowledge. The most economists need to do is take sides and let the side they choose guide their choice of problems and their choice of standards for good solutions to the problems they choose. What does seem likely is that the pretense that there is no epistemological issue with any bearing on economics will soon be dropped. When it is dropped, economists will have to face the questions that McCloskey, along with the rational expectations school, addresses: why is economic theory not very good at prediction, and why isn't it getting better? The next chapter aims to show why these questions must be addressed by economists who share the empiricist commitment to predictive improvement.

3

IS ECONOMIC THEORY PREDICTIVELY SUCCESSFUL?

In chapter 1 I held that long-term improvement in predictive success is a necessary accomplishment of any discipline that claims to provide knowledge, and especially to provide guidance to policy. I have not argued for this claim. To do so would be an exercise in epistemology, which cannot expect to hold the attention of economists. But I tried to explain why economists and scientists generally need to take sides on this issue. In chapter 2 I examined some alternatives to my claim about the strength of the economist's cognitive obligation to take sides in epistemology. Hereafter I shall assume that most economists share the commitment to improvement in predictive adequacy as a necessary condition for the certification and expansion of economic knowledge. It is widely held that neoclassical economic theory does not provide much of this sort of knowledge, and that this failure raises a serious problem, one which can be expressed in several different ways: What sort of a discipline is economics, if not a policy-relevant science? How are we to interpret economic theory if we wish both to preserve it and to absolve it of defects, in the light of its predictive poverty? Can the theory be revised or improved in ways that both retain its character and make for predictive improvement? What form should a policy-relevant predictively successful science of economic behavior take?

All these questions involve the presupposition that economic theory is predictively weak. Similarly, they assume that economic theory's predictive strength cannot be improved, that whatever weakness is found in its application is due to economic theory itself, as opposed to, say, auxiliary hypotheses required to bring it into contact with economic data. Some economists will not grant this assumption. Other economists and even more noneconomists will find this assumption obvious and beyond argument. Indeed, it is enshrined in a history of jokes about the subject: "if you put all the economists in the world end to end, they still won't come to a conclusion," etc. Where does the burden of proof lie—with those who defend the predictive attainment and prospects of the theory or with those who doubt them? In this chapter I take up the burden, and try to share it with economists who are on record as defending the predictive success of their discipline! I try to show that their methodological position would be pointless except against the background of failure to improve the predictive power of economic theory.

FRIEDMAN ON THE PREDICTIVE WEAKNESS OF ECONOMIC THEORY

This burden of proof has sometimes been admitted by the most fervent defenders of neoclassical economics. It would be difficult to understand the point of Milton Friedman's famous paper "The Methodology of Positive Economics"[1] except against the background of such an admission. At first blush it might be supposed that Friedman's views are quite adverse to doubts about the predictive merits of economic theory. Friedman endorses without qualification the claim that the goal of science is "valid and meaningful (i.e., not truistic) predictions about phenomena not yet observed" (p. 7). His aim is to vindicate economic theory as attaining this goal. But there is an important qualification in his vindication: for economic theory is deemed to have been predictively successful with respect "to the class of phenomena which it is intended to explain" (p. 8). However, the question arises, what is the intended domain of economic theory? Along with other economists, like Hicks,[2] Friedman holds the intended phenomena to be facts about markets, industries, and economies as a whole. By contrast, economics is not, ac-

1. In *Essays in Positive Economics* (Chicago: University of Chicago Press, 1953). Page references in this section are to this work.
2. Sir John Hicks, *Value and Capital*, 2d ed. (Oxford: Oxford University Press, 1946), p. 76.

cording to Friedman, about individual economic choice, and therefore the failure of its purported predictions about such phenomena do not count in assessing its predictive success. This conclusion is the whole point of his famous attack on the relevance for assessing economic theory of testing its "assumptions" about individual expectation, preference, and optimization and the boundary conditions within which they operate. It is a defense of neoclassical economics against the charge that its predictive record with respect to individual economic choice is lamentable and that the boundary conditions it stipulates are never realized. The defense consists in admitting the charge but denying its relevance to any assessment of economic theory.

Friedman's argument raises three questions. Are its claims about methodology in general sound? Are its claims about the predictive success of economic theory with regard to the allegedly "intended" domain of economic aggregates borne out? Is the denial of any concern on the part of economic theory with individual behavior sustainable, or is it special pleading?

The first question I have dealt with at length,[3] as have many others. I shall say little more about it here. Let us turn to the second question. Friedman's evident aim is to defend economic theory against the claim that its assumptions are unrealistic. But why should this issue ever have arisen? Surely, the "unrealism," the idealized character of assumptions throughout the most successful theories of physical science, has long ago settled the question of whether unrealistic assumptions are permissible in predictively successful scientific theorizing. The question of whether such assumptions are warranted in economic theory was not prompted by any general unease in the philosophy of science. It resulted from doubts about the predictive success of the theory for aggregate phenomena and the desire to locate the obstacles to such success. This much appears to be clear in the very development in economic theory that seems to have motivated "The Methodology of Positive Economics."

The theory of monopolistic and imperfect competition is one example of the neglect in economic theory of

3. *Microeconomic Laws* (Pittsburgh: University of Pittsburgh Press, 1976), pp. 155–70.

[Friedman's strictures against testing assumptions.] The development of this analysis was explicitly motivated, . . . by the belief that the assumptions of "perfect competition" or "perfect monopoly" said to underlie neoclassical economic theory are a false image of reality. And this belief was itself based almost entirely on the directly perceived descriptive inaccuracy of the assumptions rather than on any recognized contradiction of predictions derived from neoclassical economic theory. The lengthy discussion on marginal analysis in the *American Economic Review* some years ago is an even clearer . . . example. The articles on both sides of the controversy largely neglected what seems to me clearly the main issue—the conformity to experience of the implications of the marginal analysis—and concentrate on the largely irrelevant question whether businessmen do or do not in fact reach their decisions by consulting schedules, or curves, or multivariable functions showing marginal cost and marginal revenue. (P. 15)

As Blaug notes, however, "the original appeal of Chamberlain's book [*The Theory of Monopolistic Competition*] was that its predicted consequences were directly contrary to the implications of perfectly competitive models." Blaug gives an example: profit maximizers in perfect competition have no incentive to advertise. "However, advertizing expenditures in an increasing number of product markets are a well-attested phenomenon."[4] This observation is British understatement at its best! But the theory of monopolistic competition predicts that producers of a heterogeneous product will advertise. It might be argued, on Friedman's behalf, that the theory of imperfect competition arose from dissatisfaction with the Marshallian notion of the "representative firm"—surely an unrealistic idealization in the theory of perfect competition. But dissatisfaction with this notion stems, not from its unrealism, but from the fact that such firms do not enjoy increasing or decreasing returns to scale. And the existence of this phenomenon is a fact economists wanted to predict and explain. The recourse to theories of im-

4. Mark Blaug, *Economic Theory in Retrospect*, 3d ed. (Cambridge: Cambridge University Press, 1978), p. 416.

perfect competition reflected this aim and not any intrinsic hostility to unrealistic assumptions. After all, the theories offered to account for imperfect competition had enough unrealistic assumptions of their own!

Much the same must be said about Friedman's other example. It is true that the marginal theory of the firm in a perfectly competitive industry, when combined with initial conditions about changes in demand, for example, makes some definite predictions about changes in supply and price, which other theories, with more realistic assumptions about the behavior of managers, do not. But among these predictions at least half are disconfirmed, and the ones that are confirmed have a serious difficulty: economists have not seemed able to improve on them in any systematic way over the entire course of the history of the neoclassical theory of the firm. The confirmed predictions about markets and industries of this theory have at best always been "generic" or, in Samuelson's term, "qualitative." When correct, they have told us the direction of a change, but there has been no tendency to go further, from the direction of a change to a well-confirmed statement of its amount or size. We shall return to the generic nature of economic predictions below. But the point here is that dissatisfaction with "the assumptions" of perfect competition or marginalism has been a consequence of dissatisfaction with the predictions of these theories and not a source of discomfort to economists by themselves. To this extent Friedman's argument against rejecting unrealistic assumptions both attacks a straw man and begs the question: no economist has ever questioned the assumptions of neoclassical theory just because they are unrealistic, and the real reason many have done so is because of dissatisfaction with what Friedman assumes to be beyond question, the predictive success of neoclassical economic theory.

This brings us to the third question: Friedman restricts the domain of predictions that permissibly test economic theory to the domain "it is intended to explain" (p. 8). One wants to ask "intended by whom?" If Friedman or any particular economist is the authority on what the "intended" domain of economic theory is to be, then his or her thesis is altogether too easy to defend. Indeed, it becomes vacuous, for it is open to that economist to simply reject any disconfirming test of any consequence of the theory as beyond the intended domain. More seriously, a cursory examination of the history of neoclassical theory shows that the intended domain of economic explanation certainly included the very phenomena described in the assumptions of neoclassical theory. Wicksteed, for example, opens *The Common Sense of Political Economy* with an explicit account of marginal utility as a literal explanation of individual human action, both economic and otherwise. He concludes the analysis thus: "Our analysis has shown us that we administer our pecuniary resources on the same principles as those on which we conduct our lives generally. . . . in the course of our investigations we have discovered no special laws of economic life, but have gained a clearer idea of what that life is."[5] The fundamental assumptions of economic theory are psychological generalizations that explain individual choice and, by aggregation, economic phenomena.

In the course of the history of economic theory since Wicksteed and the marginalists, there has in fact been less and less emphasis on the explanation of individual choice as an aim of economic theory. But it remains to be seen whether this shift in the "intended" domain of explanation is anything more than an ad hoc restriction reflecting the disconfirmation of the theory with respect to such behavior. The shift from cardinal to ordinal utility and eventually to revealed preference certainly pruned economic theory's picture of individual choice. But this shift certainly appears to be one responding to the falsification and/or trivialization of successive versions of the theory of consumer choice. Revealed preference theory allows the economist to be almost entirely neutral on the psychological mechanism that leads to individual choice. Even this minimal approach, however, makes implicit psychological assumptions, to whit, that economic agents' tastes have not changed over the period in which we offer them pairwise choices of alternatives.

Economists may confidently announce, along with Hicks,

5. (London: Macmillan, 1910), p. 126.

that "economics is not in the end very much interested in the behavior of single individuals."[6] But this interest will not prevent false assumptions about individuals from bedeviling predictions about the economic aggregates made up of them. Similarly, thermodynamics may not be interested in individual molecules, but it is facts about these molecules that explain why the ideal gas law fails beyond a narrow range of temperature, pressure, and volume. The neglect of economic assumptions can no more be justified by restrictions on the intended domain of economic theory than can thermodynamics be insulated from atomic theory. Or at least it can't if economics makes claims to being a theory with predictive consequences.

The upshot is that so far from constituting a defense of the predictive successes of economics, Friedman's "Methodology of Positive Economics" is rather a symptom of its weaknesses, with regard to both aggregate economic phenomena and individual economic behavior.

LEONTIEF ON THE SAME SUBJECT

Friedman's argument for the predictive success of economics turns out to be much less vigorous than it appears, but perhaps the demand that economic theory reflect substantial predictive success is too strong. After all, a discipline must walk before it can run. Perhaps the most we should expect of neoclassical economic theory is improvement in predictive success. Suppose we can show that though the theory has not met with many predictive successes, it has met with some, and that the number, precision, and significance of its successes are growing over time. If so, economists would, I think, have a right to be satisfied that their theory was on the right track, and they would with some justice be able to ignore complaints as wanting in patience.

There is of course a long tradition of criticism of economic theory that denies even this minimal achievement, indeed that alleges economists are not even interested in, as it were, walking so that some day the discipline might run. Although such criticisms are often founded on a more or less intuitive survey of

6. *Value and Capital*, p. 34.

the literature of economic theory, sometimes they reflect allegedly hard data about the self-imposed insulation of economic theorists from the application of their theories to data. Perhaps the most persistent critic in this latter vein is Wassily Leontief, a Nobel laureate in economic theory whose strictures cannot be fobbed off as uninformed or splenetic.

Leontief does not argue that economic theory has made little progress in predictive power. He simply assumes it. In "Theoretical Assumptions and Nonobservable Facts," his 1970 presidential address to the American Economic Association,[7] Leontief attempted rather to explain why economic theory has shown only minimal improvement and condemned the discipline for its indifference to this fact. Leontief's assessment of the state of economic theory does not begin with any Luddite rejection of the highly mathematized character of the discipline. He does complain about the indifference to testing the assumptions of most mathematical models. He holds that

> it is precisely the empirical validity of these assumptions
> [of formal models] on which the usefulness of the entire
> exercise depends. What is needed, in most cases, is
> a[n] . . . assessment and verification of these assumptions
> in terms of observed facts. Here mathematics cannot help,
> and because of this, the interest and enthusiasm of the
> model builder suddenly begin to flag: "If you don't like my
> set of assumptions, give me another and I will gladly make
> you another model; have your pick." (P. 274)

It is no surprise that this claim should fall on deaf ears among economists. Most have adopted Friedman's attitude with respect to assumptions. Moreover, even if Friedman's attitude is too complacent, Leontief's is too harsh. Unrealistic assumptions are not problematic in general, and the mere fact that economic theory begins with idealizations cannot by itself be the source of its difficulties.

But Leontief has a better argument for the same conclusion that economic theory shows no signs of improvement. Turning

7. Reprinted in *Essays in Economics* (New Brunswick: Transaction Books, 1985). Page references in this section are to this work.

from mathematical models in the theory, he asks, "But shouldn't this harsh judgment be suspended in the face of the impressive volume of econometric work?" The answer, he writes, "is decidedly no." Advances in econometrics are in his view attempts to design statistical tools to extract significant findings from a data base that is weak and apparently not growing. What is more,

> like the economic models they are supposed to implement, the validity of these statistical tools depends itself on the acceptance of certain convenient assumptions pertaining to stochastic properties of the phenomena which the particular models are intended to explain—assumptions that can be seldom verified. In no other field of empirical enquiry has so massive and sophisticated a statistical machinery been used with such indifferent results. Nevertheless, theorists continue to turn out model after model and mathematical statisticians to devise complicated procedures one after another. (Pp. 274–75)

So, it is not just the unreality of the assumptions of economic theory that results in its lack of predictive progress; it is also the fact that economists do not or cannot augment the data that will enable them to identify progress in prediction, nor have their econometric techniques proved powerful enough to derive results about the slim body of data already in hand. And, worst of all, there is no institutional motivation in the discipline to improve on this data base: advances in formal modeling or the mathematics of statistical testing are more highly prized in economics than the augmentation of data.

One possible reply to this charge needs to be set aside immediately, though Leontief does not specifically address it. The reply is that the collection of data must be guided by theory. Without a fairly specific body of hypotheses to identify what variables are to be measured, observation will be at best haphazard and at worst pointless. Economic theory has perhaps not developed far enough to indicate what sort of data will test it adequately and what kinds of aggregate or individual phenomena it can be expected to predict. To this appeal for patience with economic theory, Leontief might reply in terms of a claim he made twenty years before he wrote "Theoretical Assumptions and Nonobserved Facts":

> If the great 19th-century physicist James Clerk Maxwell were to attend a current meeting of the American Physical Society, he might have serious difficulty in keeping track of what was going on. In the field of economics, on the other hand, his contemporary John Stuart Mill would easily take up the thread of the most advanced arguments among his 20th-century successors. Physics, applying the method of inductive reasoning from quantitatively observed events, has moved on to entirely new premises. The science of economics, in contrast, remains largely a deductive system resting upon a static set of premises, most of which were familiar to Mill and some of which date back to Adam Smith's *Wealth of Nations*.[8]

In short, economists have had the same theory in hand for upwards of two hundred years. The theory has not been increasingly accommodated to data it has led us to uncover; it has not even motivated much augmentation of data. Leontief's arguments can hardly be treated as expressions of impetuousness.

Leontief makes an important exception to his assessment of contemporary economics: agricultural economics. Here he says there has been a "healthy balance" between theory and application and a resulting secular improvement in predictive accuracy. Without gainsaying these improvements, however, two things should be noted. First, many of the problems of agriculture to which economic thinking has been applied are "technological"—they are problems of constrained maximization of production or minimization of cost where the constraints are not economic but physical and are understood in great quantitative detail through the application of the "hard" sciences. The very factors Leontief cites—crop rotation, fertilizer effects, alternative harvesting techniques—are simple enough variables to bring the economy of the farm far closer to the assumptions of neoclassical theory than the economy of

---

8. "Input-Output Economics," *Scientific American* 185 (1951): 15–21, quotation from p. 15.

other units of production. This simplicity suggests that rather than holding out promise for the rest of economic theory, the success of agricultural economics helps us see its predictive limits. Beyond the solution of linear programming problems, economic theory itself has added little to what the agricultural sciences themselves can tell us about farm production. Second, agricultural economists themselves complain in terms very like Leontief's own about the difficulty of increasing neoclassical economics' quantitative understanding of the operation of agricultural markets. Distinguished agricultural economists lament the overdevelopment of formal theory and econometric techniques and complain that the discipline does not adequately serve the needs of agriculture.

The advantage of agricultural economics is the far greater completeness of the data base deemed relevant to agricultural policy. As much as anything, the existence of these systematic and reliable data is responsible for the successes of the subject. But the data exist in large part because they are easier to secure and far simpler to organize into homogeneous categories than the data that could test economic theory elsewhere. Moreover, the hypotheses about the data that are most strongly confirmed are, as noted above, technological and not economic. One of the problems of nonagricultural economics is the difficulty of securing data. This difficulty has encouraged economists to concentrate on theory instead of applications, and so engendered Leontief's problem. To begin to correct the situation, if it can be corrected, we need the right explanation of why the provision of these data is so difficult.

Twelve years after offering his assessment of the problems of economics to the American Economic Association, Leontief's views had not changed. In "Academic Economics"[9] he summarized a content analysis of the leading journal in economics. This analysis showed that over a ten-year period the proportion of papers in the *American Economic Review* that elaborated mathematical models without bringing the models into contact with data exceeded 50 percent and that a further 22 percent involved indirect statistical inference from data previously published or

9. *Science*, 9 July 1982, p. xii.

available elsewhere. The next largest category, approximately 15 percent, was described as "analysis without mathematical formulation and data." Leontief concluded:

> Year after year economic theorists continue to produce scores of mathematical models and to explore in great detail their formal properties; and the econometricians fit algebraic functions of all possible shapes to essentially the same sets of data without being able to advance, in any perceptible way, a systematic understanding of the structure and operations of a real economy.

ARE GENERIC PREDICTIONS ENOUGH?

It is not that economic theory has no predictive power; rather the problem is that it does not have enough. And it never seems to acquire any more than it had at the hands of, say, Marshall in the late nineteenth century. We can illustrate this limitation by a glance at Paul Samuelson's *Foundations of Economic Analysis*.[10] So far as imposing demands of testability and prediction on economic theory, this work is the latter-day locus classicus. Samuelson's explicit aim was to formulate operationally meaningful theorems, by which he meant "a hypothesis about empirical data which could conceivably be refuted" (p. 4).

> In this study I attempt to show that there do exist meaningful theorems in diverse fields of economic affairs. They are not deduced from thin air or from a priori propositions of universal truth and vacuous applicability. They proceed almost wholly from two types of very general hypotheses. The first is that the conditions of equilibrium are equivalent to the maximization (minimization) of some magnitude. [The second is] the hypothesis . . . the system is in "stable" equilibrium or motion. (P. 5)

These two features, equilibrium and its stability, will be of considerable importance in subsequent discussion. What is worth noting here is that the sort of empirical tests to which Samuelson

10. (Cambridge: MIT Press, 1947). Page references in this section are to this work.

insists economic hypotheses be exposed are not quantitative but, as he calls them, "qualitative":

> In cases where the equilibrium values of our variables can be regarded as the solutions of an extremum (maximum or minimum problem), it is often possible regardless of the number of variables involved to determine unambiguously the qualitative behavior of our solution values in respect to changes of parameters. (P. 21)

But what exactly are qualitative predictions? Samuelson writes:

> In the absence of complete quantitative information concerning our equilibrium conditions, it is hoped to be able to formulate qualitative restrictions on slopes, curvatures, etc., of our equilibrium equations so as to be able to derive definite qualitative restrictions upon the responses of our system to changes in certain parameters. It is the primary purpose of this work to indicate how this is possible in a wide range of economic problems. (P. 20)

The qualitative predictions that we can test against data are roughly the signs, positive or negative, of the partial differentials of the changes in the values of economic variables we set out to measure. Thus, to take one of Samuelson's simpler examples, we may predict that an increase in the tax rate will lead a firm to a lower or higher output depending on whether the second partial derivative of output with respect to the tax rate is positive or negative. And though we cannot reliably detect the actual quantitative value of this variable, we can more easily determine its sign, that is, determine whether the first partial derivative of output with respect to taxation is increasing or decreasing.

In general, qualitative predictions purport to identify the direction in which changes move, without, however, identifying the magnitude of these changes. Of course, as Samuelson notes, we would like to have more than qualitative predictions, for as Samuelson writes,

> purely qualitative considerations cannot take us very far as soon as simple cases are left behind. Of course, if we are

willing to make more rigid assumptions, either of a qualitative or a quantitative kind, we may be able to improve
matters somewhat. Ordinarily the economist is not in pos-
session of exact quantitative knowledge of the partial
derivatives of his equilibrium conditions. None the less
[*sic*], if he is a good applied economist, he may have definite notions concerning the relative importance of different effects; the better his judgement in these matters,
the better an economist he will be. These notions, which
are anything but a priori in their original derivation, may
suggest to him the advisability of neglecting completely
certain effects as being of a second order of magnitude. . . . In the hands of a master practitioner, the
method will yield useful results; if not handled with caution and delicacy, it can easily yield nonsensical
conclusions. (Pp. 26–27)

But unless we have an independent criterion of caution and delicacy, the difference between useful results and nonsensical
conclusions will just be the difference between confirmed ones
and disconfirmed ones. In the forty years after Samuelson published *Foundations,* no such independent criterion has been
forthcoming. But without it, or without well-substantiated nu-
merical values for the parameters and variables of economic
theory, neoclassical economics is condemned to generic prediction at best.

By generic predictions I mean predictions of the existence of
a phenomenon, process, or entity, as opposed to specific predictions about its detailed character. Generic prediction is characteristic of most theories that proceed by identifying an
equilibrium position for the systems whose behavior they describe and then claim it moves toward or remains at this equilibrium value. Classical examples of such theories are to be
found in thermodynamics and evolution. (In fact Samuelson
identifies his method with that of thermodynamics [p. 21].)

Evolutionary theory, as noted in chapter 1, cannot, in the nature of the case, predict the course of evolution; at most it informs us that over the long run, biological lineages will manifest
increasing local adaptation to their environments. The theory

tells us they are moving toward an equilibrium level of population and of other individual traits that maximize their prospects for survival. For further details about the particular character of these adaptations, why they obtained instead of other equally advantageous alternatives, how they work to secure adaptation, and so on, we need to appeal to other, nongeneric, specific theories. Similarly, the second law of thermodynamics informs us that energetic systems in the long run move toward an entropy-maximizing equilibrium, without telling us what course they take toward this equilibrium level. This lack of specificity is a weakness in generic theories, but it is an unavoidable concomitant of their explanatory strategy. A theory that explains behavior by claiming that it always remains in or near some equilibrium is bound to be generic, for it must be consistent with whatever is done by the system whose behavior it describes.

In general, however, generic predictions are not enough. They are not a natural stopping place in scientific inquiry. And in other disciplines, generic theories have been either supplemented or improved in order to make specific predictions. Thus, if we add theories about heredity, physiology, development, behavior, and environment to evolution's mechanism of variation and natural selection, we can hope to increase the specificity of generic predictions. In thermodynamics, if we provide a measure of entropy and a description of the mechanical and thermal properties of a system, we can make specific predictions about the amount of entropy increase it will manifest.

This is what we need to do in economics: either supplement the theory with theories from other disciplines that will enable us to convert generic into specific predictions, or find measures of the independent or exogenous variables of the theory that will enable us to do so. We need to do these things if economics is to justify any confidence as a policy-relevant science. Generic prediction is something; it is a start, but it is not enough. And economists should not be satisfied with it.

GENERIC PREDICTIONS IN KEYNESIAN MACROECONOMICS AND RATIONAL EXPECTATIONS THEORY

The shift from classical economic theory to Keynesian macroeconomics and the rational expectations theory counterrevolution illustrates the degree to which economics as a discipline seems unable or unwilling to transcend generic predictions. The force of the illustration does not rest on adopting any of these three theories, least of all the most recent, and it is only one of many examples. Its advantages for our purposes are that it reveals in the context of lively debate in recent economic theory the role of prediction in economics and the limitations upon it.

Classical economic theory predicted the existence, stability, and uniqueness of a market-clearing general equilibrium. True, the existence was not mathematically "nailed down" until the thirties, nor was uniqueness established till even later, and we still lack a complete account of stability. Nevertheless, counting equations and unknowns seemed enough to substantiate the generic prediction of equilibrium. That the prediction was generic should be obvious. Business cycle "fluctuations" could be accommodated to the theory only insofar as its assertions were not specific but generic. Indeed, business cycle theory was itself restricted to the generic explanation of the possibility of such fluctuations and never reached the point of attempting to project their dates, durations, and magnitudes.

The depression and deflation of the 1930s seriously undermined confidence in even the weak generic predictions of classical theory. To be sure, there were ways of reconciling the existence of a market-clearing equilibrium somewhere in the long run. But, as Keynes noted, in the long run we are all dead. So, Keynes developed a theory in which a non-market-clearing equilibrium was possible, in particular one that did not clear the labor market and so explained (generically) the existence of involuntary unemployment. The possibility of such a non-market-clearing equilibrium is the interpretation of *The General Theory* taught in most contemporary macroeconomic classes.

In the textbooks Keynes's theory is presented in terms of a

set of standard equations, with real national income as a function of consumption and investment (with these in turn functions of the interest rate and income), the demand for money balances as a function of income and the interest rate, and aggregate production and the demand and supply of labor dependent on the wage rate and the price level. The consequences of the theory for fiscal and monetary policy are given in the well-known Hicks-Hansen IS-LM curve (fig. 1).

The IS curve gives all the values of national income and the interest rate that equilibrate planned investment and planned savings, whence the name IS. The LM curve reflects the set of points of income and interest rate at which the supply and demand for money balances are equal. The point at which these curves cross has the property of being a level of national income and interest rate at which planned savings and investment are equal, and there is no excess demand or supply of money. In the standard view of the shape of these curves, this point is a stable equilibrium.

Two important consequences are usually drawn from Keynesian theory: first, an explanation of how unemployment equilibria are possible, and second, a recipe for "fine-tuning" an economy to keep it close to a full-employment equilibrium level of income. The first is often expounded in terms of how the shapes of the IS and LM curves can be affected by wage rigidities, the phenomenon of the liquidity trap, and an investment demand curve that is inelastic for changes in the interest rate.

The second implication, that by fiscal and monetary policy the government can shift the IS and LM curves to the right to increase the equilibrium levels of income and thus lower the level of unemployment, proved in some ways a more exciting consequence of the theory. For one thing, Keynes's explanation of the depression as a phenomenon of unemployment equilibrium was the explanation of a possibility. That is, it showed how something conventional economic theory deemed to be impossible might after all be possible. If, as many held, such an equilibrium was not merely possible but actual, then economic theory had to be reconciled with it. In establishing how unemployment equilibria were possible, Keynes preserved economic theory from the charge of irrelevance in principle. But neither

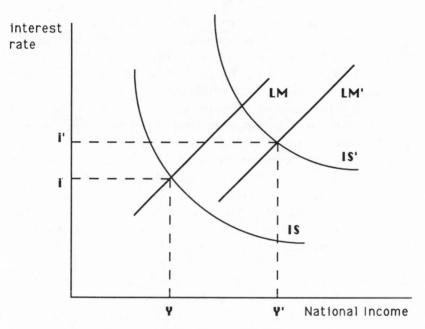

interest
rate

Figure 1. Hicks-Hansen IS-LM curve.

Keynes nor his successors did much more than show this possibility. And the test of his theory's explanatory power was taken to be its implications for stabilizing the economies of industrial nations in the period after the depression ended. Here there were hopes of doing more than merely showing that what most people deemed to be actual was after all possible. The prospects of shifting the IS and LM curves to fine-tune the economy was just what a discipline hankering for predictive success needed.

Suppose that the equilibrium value of national income, Y (on fig. 1), is below a full-employment level, and suppose that there is no budget deficit. Then by shifting the IS or the LM curve, or both, to the right, the government can raise the equilibrium level of income and thus reduce unemployment. The IS curve is shifted to the right by increasing government spending. The increased spending when added to consumption and private investment results in a new equilibrium level of income, Y', higher than Y, at the point of intersection of the LM and new IS curves. The government's new spending, however, produces a deficit, which can be made up by either selling bonds or increasing the money supply. Suppose the government sells bonds. This increase in the supply of bonds can be absorbed only at a higher interest rate. This raises interest rates above i, as the rightward shift in the IS curve indicates.

On the other hand, the government can finance the deficit by increasing the money supply instead of selling bonds. Increasing the money supply, however, shifts the LM curve to the right, thus moving the equilibrium level of national income even higher. The increase in real national income, however, increases demand and therefore reduces the real value of the money supply; that is, it produces inflation in the price level. On the other hand, the increased national income produces higher tax revenues and reduces the budget deficit. If all this happens at the right rates and in the right amounts, the end result is a balanced budget at a new and higher level of national income and a lower level of unemployment. The units on any IS-LM curve are purely notional, and the Keynesian predictions are purely generic. It is important to bear in mind that econometric models that attached "real numbers" to macroeconomic variables never unequivocally confirmed this ideal model.

Given the theory, fine-tuning was supposed to work. But it didn't. In particular, the economic phenomena of the seventies showed that the so-called Philip's curve, tracing the trade-off between unemployment and inflation that results from increasing the money supply (i.e., shifting the LM curve to the right), either broke down or did not hold at all. The generic predictions of Keynesian theory never had a chance to give rise to quantitative ones before they began breaking down. Ironically, having first shown how the actual was possible in the case of unemployment equilibrium, economics now had to show how the actual was possible in the case of "stagflation"—inflation with increases in unemployment.

At this point the rational expectations theory counterrevolution took hold. It began with a "hypothesis" of John Muth: "expectations, since they are informed predictions of future events, are essentially the same as predictions of the relevant economic theory. At the risk of confusing this purely descriptive hypothesis with a pronouncement as to what firms ought to do, we call such expectations 'rational.'"[11] The supposition that economic agents are rational in framing their expectations as well as in acting on them was most famously applied to the Keynesian model by Robert Barro.[12]

The classical theory is sometimes said to deny the possibility of an unemployment equilibrium because economic agents are rational. Keynes is believed to have proved the possibility of an unemployment equilibrium by assuming that people are not fully rational. For example, the stickiness of wages reflects the existence of a "money illusion" that the classical theory of consumer choice proscribes.

The failure of fine-tuning, which seems to disconfirm the conventional Keynesian model's generic predictions, naturally led to diagnoses and new non-Keynesian macroeconomic models. But, remarkably, these diagnoses and alternatives were in fact "throwbacks" to classical theory. Adopting Muth's "hypothesis" (rational expectations theory) does not so much represent an advance on Keynesian theory as a return to his classical predecessors.

According to Barro's rational expectations approaches to the IS-LM curve, government action cannot raise national income by shifting these curves, because people are too rational to be tricked by such policies. Or at least enough of them will know that the government embraces Keynesian theory and will have expectations about the effects of government policies, so that these policies will be ineffective.

Rational expectations theorists sketch the microfoundations for the Keynesian theory in the following terms. In order to finance the deficit resulting from increased spending, the government must either sell bonds or increase the money supply. Suppose it sells bonds. This is supposed to result in an increase in people's perceived wealth, the bonds being counted as among their assets. Because people are now richer, they may spend more on commodities and services, therefore raising the

11. "Rational Expectations and the Theory of Price Movements," *Econometrica* 29 (1961): 315–35, quotation from p. 319.
12. "Are Government Bonds Net Wealth?" *Journal of Political Economy* 82 (1974): 1095–1118.

level of national income and shifting the IS curve even further to the right. But, Barro claims, when the government sells bonds to cover the deficit it has created (by shifting the IS curve), rational agents realize that servicing this debt will require higher taxes, sooner or later. Accordingly, the alleged wealth effects of holding new government bonds are at least blunted if not completely overborne by the expectations of higher taxes. And this expectation leads to more saving and less consumption, thus lowering national income back in the direction of Y. In other words, governmental steps to increase national income by borrowing will not do so if economic agents are rational and have correct beliefs about the economic model the government employs and its policy ramifications.

Similarly for any attempt to cover the deficit through increases in the money supply, the rational agent recognizes that increasing the money supply will reduce the real value of his or her monetary holdings, by increasing the price level. Accordingly, rational agents will have to save more money in order to maintain the real value of their money balances. Increased savings means decreased consumption, thereby blunting the original effects of an increase in the money supply. In this view, neither monetary policy nor fiscal policy can be effective, because they are predicated on the false assumption that economic agents' expectations about the future are not well informed. Keynesian theory is held to assume that agents' expectations about the future are a function solely of their knowledge about the past: they simply extrapolate current prices and quantities forward into the future. This claim is misleadingly called the assumption of "adaptive expectations"—misleading because the expectations track the past and will not be adaptive if policy changes.

The rational expectations view is not that all economic agents are omniscient about the future, nor even that most agents' expectations are better than merely "adaptive." It holds merely that at least some agents employ all information available in order to formulate their expectations, not just past values of economic variables; it holds that aggregated expectations about the future are on average correct, or at least that enough agents are correct enough of the time to take economic advantage of governmental policy changes, and that the economic incentives for doing so are great enough for their actions to affect the direction of the entire economy.[13]

The most ardent proponents of rational expectations theory argue that it shows no governmental policies can attain the fine-tuning that Keynesian theory aims at. But this conclusion is far too strong. Even if it were shown that rational expectations will "outsmart" every sort of fiscal and monetary policy based on a Keynesian model, at most this would show that the Keynesian model embraced by the government was faulty, not that no macroeconomic theory of any kind is possible. Rational expectations theory is not a general proof that all macrotheories must fail through "reflexivity" (i.e., as a result of people coming to know about them).

Even among those proponents of rational expectations theory less extreme in their claims about the impotence of macroeconomic policy, the theory is a return to classical orthodoxy. The rational expectations critique of Keynesian theory is that it fails to treat people as economically rational agents, thoroughgoing optimizers, that it surrenders without warrant the classical assumption that markets all clear at equilibrium, and that it fails to link up with the well-developed body of price, value, and welfare theory. As for Keynes's original motivation, to explain the depression by explaining how an equilibrium that did not clear the labor markets was possible, it simply denies that there is any such thing:

> Involuntary unemployment is not a fact or phenomenon
> which it is the task of theorists to explain. It is, on the con-

---

13. This is, I suspect, what McCloskey means when he argues that economic prediction is impossible. For if it were possible, economists, who surely have the most rational expectations, could employ their information to get rich. Because others have similarly rational expectations, economists cannot make profitable predictions, predictions that take advantage of other agents' merely "adaptive expectations" (see McCloskey's *The Rhetoric of Economics* [Madison: University of Wisconsin Press, 1985] and *If You're So Smart* [Chicago: University of Chicago Press, 1990] and chap. 1 above). Again it is worth noting that McCloskey is in effect committed to the "generic prediction" that economists will not get rich selling or using their discipline's theories.

trary, a theoretical construct which Keynes introduced in the hope that it would be helpful in discovering a correct explanation for a genuine phenomenon: large-scale fluctuations in measured, total employment. Is it the task of modern theoretical economics to "explain" the theoretical constructs of our predecessors?[14]

And, of course, rational expectations theory provides its own account of fluctuations in the business cycle and the causes of high unemployment. A sketch of one version of the rational expectations theory of fluctuations in employment shows the degree to which argument between rival theories in economics remains at the generic level. The rational expectations model is provided as at most a caricature that renders expectable the direction that phenomena take, though not the dimensions of the phenomena.[15]

Lucas offers the following account of how fluctuations in employment may reflect rational expectations.[16] Consider how a representative agent producing a common good will respond to a change in its price. In the case of a laborer, this good is his or her work and the price is his or her wage. If the price increase is expected to be permanent, there will be no effect on the level of employment (at any rate, data from labor economics suggest this). But if the price (i.e., wage) increase is viewed as temporary, the laborer will increase his or her supply of labor, working more hours, and postponing leisure to a period after the wage rate returns to its prior level. The opposite will occur when the wage rate is thought to have moved downward temporarily: vacations will be moved forward, because the opportunity cost of leisure has declined. So, labor supply varies with expectations abut the duration of wage rate changes. Employ-

14. R. E. Lucas, "Unemployment Policy," *American Economic Review* 68 (1978): 353–57, quotation from p. 354.

15. The description of such models as "caricatures" is not mine. It is that of a distinguished economist, Hal Varian, who does not employ the label as a term of abuse but to highlight something important about economic theory, as we shall see in the next section.

16. R. E. Lucas, "Understanding Business Cycles," in *Stabilization of the Domestic and International Economy,* edited by K. Brunner and A. Meltzer (Amsterdam: North-Holland, 1977), pp. 7–30.

ment fluctuates because workers speculate on the market for leisure—buying more when the price declines and less when it rises. Small random fluctuations in the wage rate can cause considerable changes in the employment level, whence the business cycle, or at least an important part of it. The reaction of some economists, especially Keynesian, to rational expectations theory is roughly that of not knowing whether to laugh or to cry. They recognize that the theory is an attempt to return to the status quo ante Keynes and surrenders all the hard-won interventionism of the mixed economy. James Tobin is eloquent on this point:

> [Rational expectations theorists] are all inspired by faith
> that the economy can never be very far from equilibrium.
> Markets work, excess supplies and demand are eliminated,
> expectations embody the best available information,
> people always make any and all deals which would move all
> parties to preferred positions. With such faith the ortho-
> dox economists of the early 1930's could shut their eyes to
> events they knew a priori could not be happening. . . .
> Keynes might say this is where he came in.[17]

Although the dispute between Keynesians and rational expectations theorists is not one for us to adjudicate, it illustrates quite clearly the fact that two hundred years after Smith and a hundred years after Walras, economic theory is still brought to bear on economic phenomena at most generically. One reason may be that economists have still not decided in what direction the long-term facts to be explained point. As a result they cannot choose which generic theory to attempt to improve in the direction of specificity. Here Leontief's criticism seems to apply. The data so far collected to test rational expectations predictions versus Keynesian ones about, for instance, the relation of interest rates to budget deficits point in no one direction. Furthermore, too little effort seems to be devoted to improving on this data, as opposed to constructing intellectually beautiful models of possible economies.

17. "How Dead Is Keynes?" *Economic Inquiry* 15 (1977): 459–68, quotation from p. 460.

On the other hand, conventional economic theory may be intrinsically limited to generic prediction. If so, the rational expectations theorists' continued devotion to it simply reflects economic theory's continued commitment to perpetually being at most generic in its claims about the world.

It is evident that rational expectations theorists are not daunted by the "unrealism" of their models, by the fact that their most attractive explanation for the failure of a theory "more realistic" than the classical one is not just less realistic but uncompromisingly classical. This property of economic theory needs to be explained. One explanation, convenient and widely embraced, is Friedman's: the "unrealism of the assumptions" is no defect in a theory; indeed it is a requirement for real explanatory power. I think we can disregard Friedman's explanation, if only because it rests on a controversial assumption about the predictive success of economic theory. Moreover, it does not specifically explain the continuing attachment of economic theory to optimizing models.

Another explanation is that prediction, even generic prediction, is not what successive economic theories aim at. In this view, successful generic prediction is at most a necessary requirement of economic models. It is not enough, but it is a start, and the rest of what the economist wants is provided by other features of his or her models.

GIBBARD AND VARIAN ON ECONOMIC MODELS AS CARICATURES

It is certainly true that economists are interested in other things besides prediction, specific or generic. Although McCloskey is quite wrong to deny that economics has or needs any predictive pretensions, the discipline cannot be focused solely, or even largely, on this aim. For if it were, it would long ago have surrendered neoclassical theory for some other more predictively powerful one. These other apparently nonpredictive interests are explored in a paper entitled "Economic Models" by a distinguished economist, Hal Varian, and an equally distinguished philosopher, Alan Gibbard.[18] What is especially important for our purposes is that in the end, Varian and Gibbard embrace the conclusion that the best we can hope for from economic models is generic prediction after all, and that this is enough because there seems no reason to suppose that such models actually provide understanding of economic phenomena.

Gibbard and Varian ask the question, "In what ways can a model help in understanding a situation in the world when its assumptions, as applied to that situation, are false?" (p. 665). It is worth restating the point of the first section here, that the general question is not at issue. We know how and why theoretical models in physics, for example, help in understanding, even though their assumptions are false. However, if the same answer—predictive success—is offered for economic models, then those models have failed to meet the criterion.

Gibbard and Varian begin by distinguishing ideal models from descriptive ones that attempt to fully capture economic reality. It is the latter that are their concern, evidently because they are more common and more important in economic theory. Within this class there is a further distinction between approximations and caricatures: models that purport to give an approximate description of reality and those that "seek to 'give an impression' of some aspect of economic reality not by describing it directly, but rather by emphasizing—even to the point of distorting—certain selected aspects of the economic situation" (p. 665). This distinction, between approximations and caricatures, is, they say, one of degree.

Now, a model is a story with a specified structure—the story gives something like the extension of some of the predicates in the structure, but it is not a story about any particular producers or consumers or any economic situation in the world. It may, however, be applied to a situation. The result is an applied model, produced by giving the structure's predicates and quantifiers a particular extension. Only when thus interpreted can we ask how close to the truth the model is.

18. *Journal of Philosophy* 75 (1978): 664–77. Page references in this section are to this paper.

Though false, the model's assumptions are hypothesized by the economist to be close enough approximations to the truth for explanatory "purposes." This explanation takes the following form:

> First, if the assumptions of the applied model were *true*, the conclusion would be—here the proof is mathematical. Second, the assumptions in fact are sufficiently close to the truth to make the conclusions approximately true. For this no argument within the model can be given; it is rather a hypothesis, for and against which evidence can be given. One kind of evidence is evidence for the rough truth of the conclusions of the applied model; another kind is evidence for the rough truth of the assumptions.

The model has explanatory power with respect to its explanandum if we have evidence that "the conclusions of the . . . applied model were close to the truth because its assumptions were close to the truth" (p. 670).

But closeness to the truth turns out almost always to be a matter of what I have called "generic" prediction or explanation. Models are employed to explain the "central tendency" of economic behavior (p. 670); "when models are applied as approximations, few if any of the degrees of approximation are characterized numerically" (p. 672). Economists, we are told, apply models to situations in two different ways: econometrically and casually. About econometric applications Gibbard and Varian say little, partly because they note there is a well-developed methodology for them, but mainly because such models are usually severe complications of simpler casual models, specifically intended for prediction. As such they do not raise the problems of nonpredictive models. On the other hand, we have the testimony of Leontief that such econometric models do not have a good track record. So, although conceptually unproblematical, they are of arguable utility.

What Gibbard and Varian call casual applications of models are both more characteristic of economic theory and far more problematical. Thus, they are right to focus on these applications.

The goal of casual application is to explain aspects of the world that can be noticed or conjectured without explicit techniques of measurement. . . .

When economic models are used in this way to explain casually observable features of the world, it is important that one be able to grasp the explanation. Simplicity then will be a highly desirable feature of such models [and] complications may be unnecessary, since the aspects of the world the model is used to explain are not precisely measured. (P. 672)

The goal of casual application, notice, is much the same as the goal Samuelson adopted in 1947. Indeed, it is very little different from the explanatory results of Adam Smith's *Wealth of Nations*. Gibbard and Varian expound casual application in ways that make the lack of advance on Smith even more palpable, for they go on to assert that one of the most important explanatory functions of an economic model is to serve as a caricature. A model is often designed to exaggerate some feature of reality, instead of approximating it, and this is essential to its explanatory power:

An applied model that ascribed to [an assumption] its approximate place in reality might bury its effects, and for that reason, a model that is a better approximation to reality may make for a worse explanation of the role of some particular feature of reality.

If the purpose of economic models were simply to approximate reality in a tractable way, then, as techniques for dealing with models are refined and as more complex models become tractable, we should expect a tendency towards a better fit with complex reality through more and more complex models. . . . [B]ut a tendency to better approximations through more complex models is by no means the rule. (P. 673)

This passage raises many fundamental questions that Gibbard and Varian do not address. For example, what is the cognitive function of explanation if a more nearly true model may turn out to be a worse explanation? Surely, the psychological accessibility of an explanation is not an important constraint in comparison to its degree of evidential support, systematical connections, and predictive applicability. Moreover, the fact that the history of economic theory does not reveal a "tendency towards a better fit with complex reality" is a matter that cries out for explanation. Gibbard and Varian are right to say that economists prize what they call caricature models. But the reason they do so, remains unspecified. One possibility is suggested in the claim that approximating models "bury" their effects. I think this means that such models bring in so many possibly countervailing economic forces that they do not issue even in generic predictions. In this interpretation, caricatures are prized because they do provide at least such generic claims about "general tendencies."

A caricature model is one in which approximating reality is not important; rather it attempts to subsume a complex phenomenon under a simple and relatively one-sided "story." Gibbard and Varian cite an example from Samuelson: a caricature model explaining intergenerational transfers of income by assuming that the population is composed of perfectly rational persons, who live in only two periods, working and retired. They might just as well have sketched one of the rational expectations models about how people react to macroeconomic policy that figure in the previous section. About such examples, they ask, when will a caricature model be helpful in understanding a situation? The answer seems to be that a caricature will explain when the explanandum phenomena "do not depend on the details of the assumptions," that is, if they are robust (i.e., follow from several different caricatures of reality).

But the claim that robust models explain is deeply perplexing. The conventional wisdom in the philosophy of science is that a theory has little explanatory power if the phenomena it purports to explain can also be explained by a wide variety of equally plausible alternative theories. (Economic opponents of rational expectations theory will agree with this claim. The failure of Keynesian policies, on which many agree, is a robust conclusion, but it does not in their view tend to substantiate rational expectations assumptions.) Robustness is an important attribute, but presumably for the explanans in an explanation and not for the explanandum—for in most cases we already know that the explanandum obtains, and we seek to explain it. Of course the robustness of consequences might do credit to a caricature, even in the presence of other equally plausible ones with the same consequences. But this will apply only if caricatures are valued for their generic predictions, because this is what having a robust consequence usually comes to among such models. We have some evidence for this interpretation in Gibbard and Varian's statement that in the employment of a caricature, the economist's hypothesis is "that a conclusion of the applied model depicts a tendency of the situation, and that this is because the assumptions caricature features of the situation and the conclusion is robust under changes of caricature" (pp. 676–77).

A caricature, we are told, "differs from an approximation . . . not only in its simplicity and inaccuracy, but in its deliberate distortion of reality. . . . to isolate the effects of one of the factors involved in the situation, or to test for robustness under changes of caricature" (p. 676). These claims are, I concede, descriptively correct ones to make about economic theorizing. But unless there is a connection between robustness of conclusions and the understanding or explanatory power caricatures provide, they still leave open the fundamental question of how "a caricature [which] involves deliberate distortion [can] illuminate an aspect of economic life." Only in Gibbard and Varian's peroration is this issue really approached: "A caricature involves deliberate distortion to illuminate an aspect of economic life. If the uses of deliberate distortion are ignored, and the job of applied models is taken to be no more than accurate approximation under constraints of simplicity and tractability, many of the caricatures economic theorists construct will seem unsuited for their job" (p. 677). But the issue is not whether economists treat caricatures as conveying understanding; of that there is no doubt. After all, nothing is more evident than the frequency of such models in the literature. The issues are how such models convey understanding and what kind of understanding is conveyed. What are economists after by way of economic knowledge (besides the sort provided, rather inadequately, by approximation models) that makes caricatures so important? These are questions Gibbard and Varian do not specifically address, though their typology of economic models makes the questions particularly pressing. The next chapters are devoted to answering them.

4

NEOCLASSICAL
ECONOMICS AS A
RESEARCH PROGRAM

For many economists Friedman's instrumentalism remains a methodological orthodoxy. Naturally, the assumptions are false, but the falsity of assumptions is an inevitable consequence of their power to systematize economic data. Instead of viewing the fundamental claims of microeconomic theory as a body of statements capable of being true or false, followers of Friedman treat them as a set of tools, useful instruments for organizing economic observations.

Increasingly, however, economists are becoming dissatisfied with Friedman's methodology of positive economics. Philosophers would delude themselves if they believed that it was their criticisms of Friedman that awoke their colleagues in economics from dogmatic slumber. Rather, I suspect, it was the increasing dissatisfaction of policymakers with the reliability of microeconomic and macroeconomic forecasting that led to a perception of economic theory in "crisis."[1] Among economists, some responded to public criticism of economic science as a whole that their special subdiscipline was exempt, and indeed the failure of the other subdisciplines was due to an unwillingness to adopt the methods and insights of the favored subject. Thus rational expectations theorists chastened their Keynesian colleagues with not giving microfoundations enough attention.

1. See the special issue of the *Public Interest* so titled and eventually published as a separate volume, *The Crisis in Economic Theory*, edited by D. Bell and I. Kristol (New York: Basic Books, 1981), with contributions from half a dozen household names in economics. Another work with a similar theme is *Economics in Disarray*, edited by P. Wiles and G. Routh (Oxford: Basil Blackwell, 1984).

But it was with microfoundations—with the assumption of rationality and the characterization of agents' preference structures—that the methodological problems of economics began. So, those economists interested in methodology but no longer satisfied with what came to be called Friedman's "F-twist"[2] began casting about for a new way to understand the aims and methods of their discipline. Though they did not surrender the F-twist for philosophical reasons, they did turn to the philosophy of science for a new approach to understanding the nature of economic theory. In particular, many students of methodology have turned to Imre Lakatos's "methodology of scientific research programs."[3]

THE METHODOLOGY OF SCIENTIFIC RESEARCH PROGRAMS

Other philosophies of science make the individual hypothesis or at most the axiomatic system of a single theory the unit of scientific assessment and change. These philosophies have always set uncomfortable standards for economists. They raise questions about the claims of economic theory: is it true or false, analytic or synthetic, falsifiable or not, directly or indirectly testable. Or else they raise questions about the individual theories from which these individual claims are drawn: does it unify and systemize observational data and can it explain economic phenomena or predict them? These questions often devolve into questions about the truth or predictive power of the constituent claims of these theories.

But, according to Lakatos, the most important unit of science

2. "The Methodology of Positive Economics," in *Essays in Positive Economics* (Chicago: University of Chicago Press, 1953), pp. 1–31. The term "F-twist" was coined by Samuelson, who actually employed this label for Friedman's claim that theories are better the more unrealistic their assumptions, but the label has come to name Friedman's methodology as a whole. See Samuelson, "Theory and Realism—A Reply," *American Economic Review* 54 (1964): 736–40.

3. "The Methodology of Scientific Research Programmes," in *Philosophical Papers*, vol. 1 (Cambridge: Cambridge University Press, 1978), pp. 1–74. Page references to Lakatos in this chapter are to this work unless otherwise noted.

is not the individual hypothesis up for test or the theory in which it figures but the "research programme," which encompasses a succession of theories, unifying them, underwriting the methods of developing them, and protecting them from unconstructive criticism. If the succession of models and theories that characterizes the history of the discipline can be shown to be part of such a research program, then we may after all be able to employ a settled approach in order to understand the nature of economic theory.

What is a research program? According to Lakatos, it is composed of a "hard core" of fundamental assumptions treated as irrefutable by members of the program; a "positive heuristic" and a "negative heuristic" respectively prescribing and proscribing methods, principles, instruments, etc., to be employed in developing theories embodying the hard core; and a "protective belt" of auxiliary assumptions and other conceptual ploys insulating the research program as a whole (though not the theories which figure in it) from threatening anomalies. Scientific theories are formulated, revised, and superseded within the research program through the application of the hard core and the heuristic to the anomalies identified in the protective belt. Lakatos applies this picture to the assessment of scientific legitimacy, to the demarcation of science from nonscience, in the following terms:

A series of theories [a research program] is theoretically progressive . . . if each new theory has some excess empirical content over its predecessor, that is it predicts some novel, hitherto unexpected fact. . . . a theoretically progressive series of theories is also empirically progressive . . . if some of this excess empirical content is also corroborated, that is if each new theory leads us to the actual discovery of some new fact. Finally, let us call [a series of theories] progressive if it is both theoretically and empirically progressive, and degenerating if it is not. We "accept" [research programs] as "scientific" only if they are at least theoretically progressive; if they are not we reject them as "pseudoscientific." (Pp. 32–34)

It is important, especially for the economists' application of this doctrine, to note that Lakatos also says, "the methodology of scientific research programmes does not offer instant rationality. One must treat budding research programmes leniently: programmes may take decades before they get off the ground and become empirically progressive" (p. 6). What is more, degenerating research programs can become progressive again.

LAKATOSIAN CONSOLATIONS
FOR ECONOMISTS

The influence of Lakatos's methodology of scientific research programs in economics was heralded in Spiro Latsis's anthology of papers *Method and Appraisal in Economics*.[4] In this work a half dozen household names in economics attempted to apply Lakatos's paradigm to exploring one or another episode in the history of twentieth-century economics. Subsequently there followed a spate of journal articles elaborating this strategy, of which perhaps the most thoughtful was D. Wade Hand's "Methodology of Economic Research Programs."[5] It is worth noting that by 1985 Hands had repented of the view that the Lakatosian approach could shed much light on economic theory.[6]

Lakatos's approach was given further and more prominent currency among economists in Mark Blaug's *The Methodology of Economics*.[7] Not only does Blaug give Lakatos considerable space in his sketch of the history of the philosophy of science, but he also endorses the employment of Lakatos's view in the

4. (Cambridge: Cambridge University Press, 1976).

5. *Philosophy of Social Science* 9 (1979): 293–303.

6. "Second thoughts on Lakatos," *History of Political Economy* 17 (1985): 1–16. Other papers examining and/or advocating the Lakatosian approach include, for instance, E. K. Brown, "The Neoclassical and Post-Keynesian Research Programs," *Review of Sociology and Economics*, 1981; G. Fulton, "Research Programs in Economics," *History of Political Economy* 16 (1984): 187–206; R. Cross, "The Duhem-Quine Thesis, Lakatos, and the Appraisal of Theories in Macroeconomics," *Economic Journal* 92 (1982): 320–40; idem, "Monetarism and Duhem's Thesis," in *Economics in Disarray*, pp. 78–100; and Sir Lionel Robbins, "On Latsis' *Method and Appraisal in Economics:* A Review Essay," *Philosophy of Social Science* 17 (1979): 996–1004.

7. (Cambridge: Cambridge University Press, 1980).

assessment of varying theoretical developments in economics and adopts the language of SRP (scientific research programs) in his own treatment of the subject. It should be added that Blaug's general conclusion about Lakatosianism as a philosophy of science is not wholly positive: "It is clear that Lakatos's effort to divorce appraisal from recommendation, to retain a critical methodology of science that is frankly normative, but which nevertheless is capable of serving as the basis of a research program in the history of science, must be judged either a severely qualified success or else a failure, albeit a magnificent failure" (p. 40).

The most sustained argument for viewing neoclassical economics as a scientific research program in the Lakatosian mold is to be found in the work of E. Roy Weintraub. First in *Microfoundations* and then in *General Equilibrium Analysis: Studies in Appraisal,* Weintraub has most fully developed the Lakatosian approach to economics.[8] The word "appraisal," which figures in Weintraub's title, bulks large in the applications to which economists have put Lakatos. Their aim has been to show that by Lakatosian standards for scientific theory economics comes off with a favorable assessment. Research programs are "appraised" as either progressive or degenerating, in Lakatos's terms, depending on whether they predict novel facts or not. It certainly seems to have been Weintraub's aim to show that economics should be appraised as "progressive" and therefore as scientific. A less tendentious project would remain silent on the question of "appraisal." Assessment, evaluation, and appraisal would be beyond the terms of an attempt simply to understand what the actual aims of economics are, as opposed to an attempt to show that economics satisfies standards Lakatos laid down for a successful scientific research program. As we shall see, this matter is crucial to assessing Weintraub's treatment of economic theory.

Weintraub's work is particularly important because of its ex-

8. *Microfoundations* (Cambridge: Cambridge University Press, 1979); *General Equilibrium Analysis: Studies in Appraisal* (Cambridge: Cambridge University Press, 1985). Further page references in this chapter are to the 1985 work, unless otherwise noted.

plicit focus on the appraisal of general equilibrium theory. Prior examinations of general equilibrium theory by economists and philosophers provide a foil for Weintraub's analysis. Among philosophers Weintraub subjects to closest scrutiny and sharpest criticism the views I advanced in "If Economics Isn't Science, What Is It?"[9] The view attributed to me is, roughly, that general equilibrium theory is the theoretical core of economics and that economists are only interested in new axiomatic elaborations of this theory. Furthermore, I am said to hold that economists' indifference to empirical testing makes economics, or at any rate neoclassical microeconomic theory, a species of applied mathematics and not a descriptive science of human behavior and its aggregation. In this chapter I shall treat Weintraub's sketch of my view as a first approximation to the view I will elaborate in chapter 8. One reason I do so is that the sketch does accurately enough represent the interpretation given these views by economists who have dealt with them. A second reason is that this claim that economic theory is applied mathematics provides the foil for Weintraub's development of "the methodology of economic research programs."

Weintraub writes:

> The arguments of . . . Rosenberg are only partly correct. He has fairly characterized the activity associated with creating extended interpretations of the hard core of the program [of general equilibrium analysis] as a mathematical activity. He has accurately represented the sequence of papers on existence of equilibrium as a kind of applied mathematics. What he has failed to notice is that those activities form only a part of the program of neo-Walrasian economics. Examining only the hard core of the program he has criticized the program for not being empirical; he should instead have been looking in the protective belt to see the activity of corroboration and falsification—of improvement of theories. In confusing the axiom structure with the interpreted theorems, he has looked at the lemma trees and missed the programmatic forest. (P. 119)

9. *Philosophical Forum* 14 (1983): 296–314.

According to Weintraub, economics is dominated by the "neo-Walrasian program." Its hard core consists of the following propositions: (hc1) there exist economic agents; (hc2) agents have preferences concerning outcomes; (hc3) agents independently optimize subject to constraints; (hc4) choices are made in interrelated markets; (hc5) agents have full relevant knowledge; (hc6) observable economic outcomes are coordinated, so they must be discussed with reference to equilibrium states. The positive and negative heuristics of the program are injunctions to (ph1) go forth and construct theories in which economic agents optimize; (ph2) construct theories that make predictions about changes in equilibrium states; (nh1) do not construct theories in which irrational behavior plays any role;[10] (nh2) do not construct theories in which equilibrium has no meaning; and (nh3) do not test the hard core propositions (p. 109).[11]

The protective belt of the neo-Walrasian program contains theories derived from the application of the heuristic to the hard core.

Theories in the protective belt of the program, theories developed out of the hard core by the heuristic, are appraised

10. This requirement is somewhat at odds with certain theorems in human capital theory, which Weintraub certainly counts as well within the neo-Walrasian program. These theorems about the shape of demand curves are shown to follow from explicit assumptions about irrationality. See G. Becker, *The Economic Approach to Human Behavior* (Chicago: University of Chicago Press, 1976), chap. 8.

11. It is worth noting that other writers attempting to fit neoclassical economic theory into the Lakatosian framework have differed with Weintraub and one another about the components of the theory's hard core. Latsis, who first broached the idea of a hard core, identified four components, only two of which are accepted by Weintraub; Blaug treats the hard core of neoclassical microeconomics as embodied by "rational economic calculation, constant tastes, independence of decision making, perfect knowledge, perfect certainty, perfect mobility of factors" (in Spiro Latsis, ed., *Method and Appraisal in Economics* [Cambridge: Cambridge University Press, 1976], p. 161). DeMarchi (in ibid.) claims that the interdependence of prices is a part of the hard core. I find Weintraub's list more convincing than others, but as Hausman notes, it is quite weak (*The Separate and Inexact Science of Economics* [Cambridge: Cambridge University Press, 1991]).

by the method that is appropriate for any empirical science. Thus, it is entirely reasonable to ask that demand theory, or production theory, the theory of the household labor supply, be evaluated according to Popperian methods of sophisticated falsificationism. It is sensible to ask whether the theory of black-white earnings differentials is progressive. Has it been theoretically progressive in the sense that successive variants have explained the corroborated content of the predecessor? Has there been excess content in the sense that successor theories have made new predictions? (P. 120)

By contrast, the hard core is not to be so evaluated. Its appraisal follows the criteria laid down in Lakatos's examination of method in mathematics: "Proofs and Refutations."[12] In fact the mathematical development of the neo-Walrasian program, which Weintraub accuses others of mistaking for the whole of neoclassical economics, is of relatively recent vintage. Unlike others, including the present author, who trace the establishment of the hard core of economic theory back to Walras, Weintraub sketches a history which results in the existence of the hard core of neo-Walrasianism only from the early fifties, as the culmination of a series of papers beginning with Schlesinger, Wald, von Neumann, Koopmans, Arrow, Debreu, and McKenzie (p. 113). This sequence of papers constitutes, according to Weintraub, the hardening of the hard core. By proving and by refining the proof of the existence of a general equilibrium, given standard maximizing assumptions, these economists established that the hard core of the neo-Walrasian program was "no longer problematic." Although, as Weintraub notes, "the process whereby a hard core hardens is apt . . . to bear some superficial resemblances to the activities . . . characterizing a 'degenerating' research program" (p. 112), in fact it is theoretically, if not empirically, progressive:

> Progress, for this sequence, is a sequence of interpretations of the undefined terms of the hard core such that (1) each successive interpretation is manifested in a consistent

12. *British Journal for the Philosophy of Science* 14 (1963): 1–117.

model, (2) each successive interpretation contains the interpretation of the predecessor, and (3) each allows a concept uninterpreted by that predecessor to be interpreted.

. . . The theorems in the sequences of papers . . . , which have been seized upon by economists and philosophers alike as a testament to the unscientific character of economics, instead represent a natural progression in the development of any scientific research program. . . .

What seems like "more and more of the same old model" is rather a set of interpretative extensions of the terms of the programmatic hard core. What appears to be nonchalance about the empirical content of the existence-proof models is instead a sensible division of labor. (Pp. 117–18)

Once established, proofs of equilibrium of sufficient sophistication underwrite a line of theorizing, in the protective belt, and generate a sequence of problem shifts that constitute scientific progress.

The resulting series of models has an organic unity because the models are all constructed according to rules, heuristics, that show how the hard core of the program may be developed into potentially falsifiable theories. Appraising work in this area thus requires that one identify the sequence M of models or theories and show how each M_t, compared with its predecessor M_{t-1}, does or does not predict a novel fact. Comparing M_t with M_{t-1}, does M_t have excess empirical content, some of which was corroborated? Are M_{t-1} and M_t linked by shared features drawn from the hard core of the program? (P. 110)

Thus, according to Weintraub, there is a double standard for appraising economics: one is Lakatos's yardstick for being a progressive or degenerating program in mathematics; the other is his yardstick for being a progressive or degenerating program in empirical science. When these criteria are accepted as the appropriate ones for economics, then both its tissue of Friedmanite and post-Friedmanite rationales and the collection of philosophers' doubts about whether economics is a science will be seen for what they really are: in the first case superfluous and in the second case ignorant. Friedmanite rationalization is superfluous because it attempts to justify the hard core, which is beyond criticism, and therefore beyond defense as well. Philosophers' doubts are misinformed because they ignore the important successes in the protective belt, successes generated by the hard core, which provide economics its progressive character and therefore scientific standing: "Rosenberg correctly identified the quasi-mathematical nature of progress in one area of work in general equilibrium analysis." (p. 121). Yet, claims Weintraub, I mistook the whole for one of its parts. The diagnosis offered is that Rosenberg "misinterprets applications of the theory in actual cases. . . . The market pays for the predictions of economists in ways that Rosenberg does not appreciate.[13] Nor does Rosenberg's attack take cognizance of the actual predictions on matters as diverse as the demand for electricity, the funding of social security, the deregulation of ethical drugs, and the design of reenlistment pay schedules for the armed forces" (p. 53).

RESEARCH PROGRAMS AND THE DEMARCATION PROBLEM

Something can now be seen to have gone seriously wrong in the debate about the proper appraisal of economics. If Weintraub's double standard applies, then the models and theories in the protective belt of the neo-Walrasian program are unproblematical, by any empirical standard, by Lakatos's or Popper's or Hempel's for that matter. They are "empirically progressive": successive models and theories are more and more well confirmed. And because they participate in a research program along with these well-confirmed models, the assumptions of the hard core are unimpeachably scientific. But the original debate,

13. This is a strikingly economistical way to refute a philosopher, one that Weintraub quotes from Wade Hands, "What Economics Is Not: An Economist's Response to Rosenberg," *Philosophy of Science* 51 (1984): 498. See also A. Rosenberg, "What Rosenberg's Philosophy of Economics Is Not," *Philosophy of Science* 53 (1986): 127–32, and chap. 8 below.

in which Friedman's salvo was most resounding to economists, circled around the adequacy of these models and theories in the protective belt. Friedman held that important parts of economic theory were fundamentally sound and well confirmed, particularly in static monetary theory and price theory, which he says "reaches almost its present form in Marshall's *Principles of Economics*."[14] Because he believed these theories to be empirically well confirmed, Friedman attacked the notion that the unrealism of their assumptions should undermine our confidence in the theories to which they give rise. By contrast, those who have challenged the assumptions of neoclassical theory both before and since Friedman's paper, like Leontief,[15] did so in large part because they found the predictive power and the degree of confirmation of economic theory wanting and sought the source of these failures in the unrealism of the assumptions. Every reasonably sophisticated commentator on the state of economic science is enough of an instrumentalist to allow the employment of idealizing assumptions, if their payoff for persistent improvement in the predictive accuracy of economic theory is evident. But since many view this improvement as far from evident, they find Friedman's defense of the assumptions of the hard core unavailing.

Now, Weintraub makes it clear that he shares Friedman's confidence in the successes of the neo-Walrasian program. The theory of demand, production theory, human capital theory, all these theories are, in Lakatos's terms, empirically progressive. But if they were unarguably progressive, then the elaborate defense of most of economics as a Lakatosian science after all would be superfluous. At worst, without the application of Lakatosian insights we might have been right about economics for the wrong reasons, viewing it as simply a series of hypothetico-deductive systems increasingly confirmed by their predictions. This appraisal might be too crude, but it would be correct.

However, as I tried to show in chapter 2, there is a profound

14. Friedman, "Methodology of Positive Economics," p. 41.

15. *Essays in Economics* (New Brunswick: Transaction Books, 1985). See the passages quoted from this volume in the previous chapter.

debate about the success of the theories in the protective belt of neo-Walrasian economics. It is easy to pile up Nobel laureates on either side of the question of whether economics has met the test of empirical progress. One side will include Leontief and Herbert Simon,[16] holding that traditional economic theory leaves much wanting; Weintraub's side will include Samuelson, Friedman, Debreu, and others. At this point the argument may take on an unproductive "yes it does, no it doesn't" character, since a manageable number of examples of success will not convince the doubters, and no argument short of an impossibility proof will convince the defenders of the traditional approach.

Moreover, the parties to this dispute do not share a common criterion of predictive power or empirical confirmation, because there is none. It is presumably to circumvent this fruitless controversy about results in the protective belt that someone would have recourse to Lakatosian resources for the description of science. Since this view makes far less stringent immediate demands on testability, Lakatosianism enables us to table the question of predictive success in our appraisal of economics.

The philosophers and economists whom Weintraub attacks would have no qualms abut the claims of economics to scientific status if, in their view, it manifested the features of progressiveness Lakatos demands. Accordingly, pointing to controversial successes in the protective belt of neo-Walrasian economics settles little in the appraisal of economics. Among doubtful commentators on economics the diagnosis of the empirical failures of economics is often sought among those principles that Weintraub identifies as the hard core of the program.[17] It is to block this sort of fishing expedition that Lakatos's philosophy of science can be employed, for if nothing within the research program of a science should undercut the force of statements in its hard core, then the unrealism or the unfalsifiability of the hard

16. For a recent example of Simon's long-standing critique, see "Rationality in Psychology and Economics," in *Rational Choice*, edited by R. Hogarth and M. Reder (Chicago: University of Chicago Press, 1986), pp. 25–40.

17. For instance, I traced problems in Becker's human capital theory to these sources in "Can Economic Theory Explain Everything?" *Philosophy of Social Science* 9 (1979): 509–29.

core of neo-Walrasian economics in and of itself casts no light on the proper appraisal of the enterprise it motivates.

What is more, if the neo-Walrasian program is as young as Weintraub claims, dating from no earlier than the early fifties, then it is further insulated from assessment by Lakatos's dictum that "one must treat budding research programs leniently: programmes may take decades before they get off the ground and become empirically progressive" (p. 6). Others, however, date the current research program in economics from 1874, when Walras's *Elements of Pure Economy* was published, instead of from 1953, when the theorem Walras assumed to be true was finally proved to the satisfaction of the most elegant mathematical standards. Forty years is not a long time in the life of a scientific research program, but a hundred years is not a short time either.

If no agreement on the empirical progressiveness of the neo-Walrasian program is forthcoming and if the program's components fit so neatly into Lakatosian categories and if the program is young enough for us not to make very strong demands on it, then the near-term question of appraising economics comes down to the question of whether satisfying the Lakatosian paradigm is necessary or sufficient or even generally indicative of a discipline's being a science. I call this the near-term question, in contrast to the long-term question of appraising the empirical progressiveness of this new subject. To quote an economist, in the long run, of course, we are all dead, even if we can hope to agree on what the long term shows. So the only question we can face immediately is the question of whether the Lakatosian picture solves the demarcation problem: whether it provides anything like a litmus test that will enable us to distinguish what is science from what isn't. It was certainly to solve this problem that Lakatos advanced the methodology of scientific research programs.[18] If it does not do so, it will be of little help in our attempt to understand the aims and methods of economics, or at least our attempt to show these are aims and methods characteristic of a science, in the usual sense of that term.

18. See Lakatos, "Methodology of Scientific Research Programmes," pp. 6–7, where this motivation is explicitly stated.

Now it seems quite clear that Lakatos's methodology of scientific research programs does not solve this demarcation problem. First of all it helps itself to all those methodological notions its predecessors employed, and fell afoul of. And second, because of this fact, the recipe it prescribes can be as easily satisfied by research programs in such paradigmatically nonscientific disciplines as literary theory, painting, or music. If these two things are true of the methodology of scientific research programs, then satisfying its strictures reveals nothing that we want to know about what kind of an activity economics is.

According to Lakatos, a program is theoretically progressive if "each new theory has some excess empirical content over its predecessor, that is it predicts some novel, hitherto unexpected fact." It is empirically progressive "if some of this excess empirical content is also corroborated, that is if each new theory leads us to the actual discovery of some new fact" (p. 32). Scientific programs are those which meet this criterion and are "progressive." Nonscientific (and pseudoscientific) programs are those which do not, and are therefore "degenerating." But the notions of empirical content, novel fact, and corroboration are among the most vexed ones in the philosophy of science. By and large philosophers have given up trying to make sense of them in a way that will shed any light on the problem of demarcation. The failure of a long sequence of principles of verification and falsification to adequately distinguish science from nonscience was accepted—just because these notions proved to be essentially recalcitrant to analysis. If they had not been, then the demarcation criteria based on them would have succeeded. There would have been no need for still another proposal, especially one as liberal as Lakatos's, which makes the unit of scientific appraisal a research program that may endure for centuries.

If we cannot separate out the empirical content of a theory from its logical form, as Quine pointed out in "Two Dogmas of Empiricism"[19] about the time the neo-Walrasian hard core hardened, and if corroboration is sometimes much more and

19. In *From a Logical Point of View* (Cambridge: Harvard University Press, 1953).

sometimes much less than prediction of new facts, as Goodman, Hempel,[20] and a host of others showed at about the same time, then Lakatos's demarcation principle can be no more decisively and uncontroversially applied than any of its predecessors. Given the obvious senses of these terms, such as "empirical content" and "corroboration," it will be easy for able philosophers with time on their hands to cook up research programs, replete with hard cores, heuristics, and belts, that satisfy Lakatos's dictum, even though they constitute paradigm pseudoscience, in just the ways philosophers contrived counterexamples to the positivist criterion of cognitive significance.

Since we cannot specify a sense of empirical content or a sense of corroboration distinctive of scientific theories as opposed to intellectual enterprises outside science, it turns out that Lakatosian criteria are as easily satisfied in these nonscientific disciplines as in science. One tip-off that this must be the case is to be found in the fact that Lakatos developed the methodology of scientific research programs out of the recipe for progress in pure mathematics he elaborated in "Proofs and Refutations." The only substantial difference between progress and degeneration in mathematics and empirical science turns out to be that the claims of the latter have empirical content and can be judged to be empirically progressive or degenerating, instead of merely theoretically progressive or degenerating. But if, as is widely held, mathematical knowledge does not differ from nonmathematical knowledge in being a priori, then this difference between them breaks down. Accordingly, showing that the neo-Walrasian paradigm is a Lakatosian research program is no refutation of the claim that economics is mainly an exercise in applied mathematics. (By the same token, of course, the claim that it is applied mathematics will not constitute a criticism of the discipline. It will help to classify the discipline in order to better understand its aims and standards of adequacy.)

What seemed most clearly missing from *General Equilibrium Analysis* was anything like an argument for the empirical pro-

20. N. Goodman, *Fact, Fiction and Forecast* (Indianapolis: Bobbs-Merril, 1955); Carl Hempel, "Studies in the Logic of Confirmation," *Mind* 54 (1945): 1–26.

gressiveness of the neo-Walrasian research program. Of course, if the account of generic prediction in chapter 3 is right, and if the entire motivation for appeal to Lakatos is the absence of strongly confirming empirical data, then this is no surprise. In subsequent work, Weintraub has attempted to make up this alleged omission from his Lakatosian account. But the evidence Weintraub cited in his later work for the empirical progressiveness of the neo-Walrasian program does not really discourage this view of the hard core of the program as a branch of applied mathematics. In "The Neo-Walrasian Program Is Empirically Progressive"[21] Weintraub summarized a series of three papers on the economic behavior of households that he claimed exemplified the sort of progress characteristic of a successful research program. Even leaving aside the problems surrounding the notions of testing and confirmation, the sequence reflects the limits on economic prediction and explanation.

The papers Weintraub describes reflect the application of bargaining theory to provide a model of how household membership affects an individual's demand function. The first paper assumes that instead of behaving as a unit, the household is composed of spouses who engage in bargaining to a Nash equilibrium to determine the household demand function. One conclusion drawn from this study is that given rational agents in a family, say a husband and wife, with different preferences, the household's demand for any given commodity will vary differently with equal changes in nonwage income. In other words, as the amount of leisure a husband or wife consumes changes, the household will purchase different goods. The second paper uses fifteen-year-old survey data to construct samples of married households to test the theory advanced in the first paper. Weintraub cites the authors' conclusions that the data do not disconfirm their hypothesis about household purchases differing with differences in male/female nonwage income. The authors also canvass ways in which their statistical tests of their hypothesis may be inadequate. In the third paper one of the authors extends the bargaining theory account to the joint deter-

21. In Neil DeMarchi, ed., *The Popperian Legacy in Economics* (Cambridge: Cambridge University Press, 1988).

mination of labor supply and consumption, given household membership. The author goes on to show that the model employed can account for the well-known fact that teenagers' work decisions differ depending on whether they live at home or not. Weintraub concludes:

> Instead of the usual models, which find no relation between household membership and work, [the author] finds that "except in special cases, market work and household membership are jointly determined." . . . The [bargaining] theory allows interrelated decision analysis, and "the interactions of these decisions are especially important for (1) youths of both sexes, (2) women (where marital status replaces household membership), and (3) even prime-age males who, when facing temporarily low wage rates, are ceteris paribus in a much better position to substitute 'leisure' for income when married to a worker." The major interpretation of the analysis is that "the family provides nonemployment insurance to the son: parents insure their son a minimal level of utility when he faces poor market opportunities."[22]

It is hard to know how much importance to attach to Weintraub's case. Is it the best one he could find? Is it typical and used for its illustrative purposes? Probably it represents a sequence Weintraub knows well, because the papers were written by graduate students from his university. Of course the obviousness of the consequences of the theory drawn and tested is evident and will be seized upon by the uncharitable to show how little even a supposed case of empirical progress can provide. Perhaps a more serious weakness is the generic character of the consequences and the absence of any hope that the theory can be improved upon. That is, not only are the claims made on the basis of the theory ones we knew to be true before their derivation, but the theory offers no hope of converting the claim that the phenomenon in question obtains into claims about the extent to which it obtains: quantitative predictions for individuals and aggregates about exactly how much the de-

22. Ibid., p. 218.

mand for a good varies for a $1.00/hour increase in the wage income of a woman versus a man. Note also that the improvement in empirical content, if such there be, does not involve any change in the fundamental assumptions about individual rationality—preferences and expectations. Rather, it reflects the accommodation of an obvious fact—that individual agents marry and combine their incomes—which had hitherto not been accommodated to the theory. The obviousness of the fact accommodated substantially reduces the interest of Weintraub's example because any improvement cannot be credited to a change in the distinctive assumptions of the theory about the underlying mechanism of rational choice.

Could Weintraub have made a better case for the empirical progressiveness of the neo-Walrasian program? Suppose he had argued that there has been substantial improvement in the range of generic predictions that economic theory underwrites. That is, economists have over the decades increasingly applied the neo-Walrasian program to apparently noneconomic choices and also to more and more specific contexts of economic choice. And they have extracted well-confirmed predictions about the existence, if not the quantitative dimensions, of a variety of aggregate effects of these choices.

One economist who has made something of a career of extending the "neo-Walrasian program" to new applications in economics and beyond is George Akerlof. His application of standard tools to markets with asymmetries of information shows why, among rational agents, the asymmetries in information about the reliability of used cars reduce the demand and thus the price of available cars very substantially, why similar asymmetries in information make it difficult for the healthy elderly to purchase private health insurance, why some members of minorities are employed at wages that do not reflect their qualifications, and why brand names function to reduce uncertainty about the quality of products and so command higher prices.[23] Similarly, Akerlof has provided economic accounts of phenomena more usually explained by appeal to social and

23. "The Market for 'Lemons': Quality Uncertainty and the Market Mechanism," *Quarterly Journal of Economics* 84 (1970): 488–500.

psychological factors, such as the force of mores and norms, or the phenomena described as "cognitive dissonance." All of Akerlof's applications consist in demonstrations that the actual is possible: that some phenomena whose occurrence is already well established could have resulted from the operation of rational choices. These applications do not transcend the limits of generic prediction. But they do expand the range of economic explanation.

A more systematic example of increasing the range of generic prediction is to be found in the work of Gary Becker. Much of chapter 6 is devoted to examining his version of the neo-Walrasian program, but consider its relevance in the present connection. Becker has applied the mechanism of budget-constrained preference maximization to a wide variety of phenomena hitherto untouched by economists. For instance, his *Treatise on the Family*[24] shows how polygamy and monogamy can be explained as the results of differences in a market for marriages generated by differences in productive efficiency between males and females; how differences between rural and urban fertility levels can be derived from the assumption that the number of children is rationally chosen as a function of income, the cost of having and raising children, and the higher income contribution of a smaller number of children whose upbringing and education are costly; and how maximizing utility under conditions of imperfect information about the marriage market and the features of potential spouses explains divorce and remarriage statistics. In fact, Becker applies the same theory to the explanation of mating and family structure in nonhuman species and to the evolution of family size and structure as a function of the spread of markets for more and more of the commodities families provide in traditional societies. The result is a tour de force of extending the range of utility maximization theory. But as with Akerlof, what is striking about this work is its generic character: Becker shows how the existence of a wide variety of actual phenomena could possibly have emerged as the result of conscious constrained maximization. Like most economists Becker makes no effort to establish

24. 2d ed. (Cambridge: Harvard University Press, 1989).

that any of these phenomena emerge from the aggregation of individual rational choice, and he would doubtless hold that the issue is irrelevant to the merits of the theory.

In addition to expansion in the range of generic prediction, some might hold that there has been improvement in specific prediction. For example, consider the development of risk measures for securities traded on stock markets. Over the last twenty years, economists and security analysts have learned to calculate the relative riskiness of a security compared with the riskiness of the stock market as a whole, the so-called beta of a stock. When the beta of a stock is greater than 1, its price is more volatile than the price of an average stock: if the average share changes value by say 5 percent, a high-beta share will change value by a higher percentage; if the beta is less than 1, the stock is less volatile than the average share. Once we have defined this measure of relative risk, we can derive as a condition for equilibrium among rational investors in a securities market that all securities should have the same risk-adjusted rate of return. As we all know, of course the securities markets are not in equilibrium, and the claims that they are sometimes, usually, or always near it are all controversial. (The role of equilibrium in economic theory is treated at length in chapter 7 below.) But as a device for predicting the behavior of investors, even quantitative estimates of betas are of little use. An accurate estimate of the beta of a stock does not even enable us to make generic predictions about actual securities markets beyond the chance-level of accuracy. There are many competing explanations for this limitation on our ability to predict behavior in stock markets, and in the next chapter I explore what I think are their fundamental sources.

At most, the defender of 'the neo-Walrasian program can claim an expansion in the range of generic prediction. From predicting that the imposition of a value-added tax will raise prices and reduce demand, to the insight that increasing a fine will reduce the frequency of an infraction, to the claim that there will be a change in the supply of babies for adoption as price ceilings on babies are lifted, the domain of microeconomic prediction has certainly increased. But this increased range of generic prediction is not much of a basis on which to proclaim the empirical progressiveness of the theory. To begin with, what is new about these generic predictions can be traced to the boundary or initial conditions in which the traditional theory is applied. There are no significant changes in the theory itself to which these new generic predictions can be credited. These new predictions are not the result of adding new explanatory variables to the theory or of measuring its fundamental parameters more accurately. In fact, the remarkable thing about the increased range of generic predictions is our conviction that in most of these areas behavior is not a consequence of rational choice, and yet the assumption that it is seems to accommodate the data. This is a mystery that needs to be explained, not a reason for according increased evidential strength to the theory's central assumptions. Not every expansion in the range of generic prediction can be credited to progress in improving the theory itself.

Even if we grant that the range of generic predictions has expanded, our problem of explaining why the theory has not broken through to improvements in specific prediction remains. And the role of such improvements in the assessment of the progressiveness of a research program is very great indeed—great enough to cast doubt on the claim that the neo-Walrasian program is empirically progressive.

In any case, as we shall see in the next section, the strength of Weintraub's case study of the empirical progressiveness of the neo-Walrasian program was not sufficient to prevent his eventually surrendering the whole project of establishing the scientific credentials of economic theory.

DOWN THE SLIPPERY SLOPE
TO McCLOSKEY

It was Thomas Kuhn[25] who recognized that, shorn of a reliable mark of empirical content and a coherent theory of corroboration, the natural sciences could not be said to evince progress in any sense different from that in which nonscientific activi-

25. *Structure of Scientific Revolutions* (Chicago: University of Chicago Press, 1962), see especially chap. 13, pp. 162ff, 172–73.

ties, like art or music or theology, progress. Programs, or paradigms, whether scientific or nonscientific, involve a sequence of "theories" that face and solve puzzles, until they come upon anomalies that do not submit. At these points they may be said to degenerate and are ripe for replacement. But the anomalies that scientific research programs face cannot be uncontroversially distinguished from those that other disciplines face, so that progress in all of them comes to the same thing and is measured only "internally," within the research program itself. The Archimedean point of the ideal observer that Lakatos needs as much as falsificationists and verificationists just doesn't seem to be available.

In some ways the most damaging criticism of the methodology of scientific research programs is its endorsement by Paul Feyerabend. Lakatos, he says, "arrives at a result that is almost identical to mine."[26] Feyerabend's result is methodological anarchism, "anything goes"; there is no such thing as scientific method, still less a basis for the appraisal of science or its distinction from any other activity. Feyerabend quotes Lakatos's admission that in his scheme one "may rationally stick to a degenerating programme until it is overtaken by a rival, and even after." The reason is that what is rational is always internal to the research program. Rationality of method is dictated by the hard core and expressed in the heuristic. Either the research program dictates the nature of empirical content and the standards of corroboration, in which case none is open to any kind of appraisal and methodological anarchism reigns; or else Lakatos is committed to a transprogrammatical standard, in which case his proposal must suffer the fate of other criteria of demarcation. I fear that Feyerabend is right when he says, "Lakatos' philosophy, his anarchism in disguise, is a splendid Trojan horse that can be used to smuggle real, straightforward, 'honest' . . . anarchism into the minds of our most dedicated rationalist. And once they discover that they have been had . . . they will agree that argument is nothing but a subtle and most effective way of paralyzing a trusting opponent" (p.

26. *Against Method* (London: New Left Books/Verso, 1975), p. 182. Page references to Feyerabend in this section are to this work.

200). The methodology of scientific research programs has many attractions for the economist who aims to describe the last hundred years of the history of economics. It may be useful for understanding the rise of marginalism, the Keynesian revolution, and the rational expectations counterrevolution. But it is of little use in what has come to be called the "appraisal" of economic theory, the assessment of its scientific status and achievement.[27]

In a backhanded kind of way Weintraub has come to agree with this conclusion. Despite the subtitle, *Studies in Appraisal,* of *General Equilibrium Analysis,* he adopted a very different interpretation of its aims:

> The issue of what is and what is not a science is of little concern to me either now or when I wrote the book on general equilibrium analysis that Rosenberg takes as the text for his discussion. Economists are not unsophisticated as to think that calling economics a "science" says anything about what economists do or should do.
>
> For myself I adopted the MSRP framework to tell a story of the development of an area of economics, and to exhibit a variety of connections among areas in economics that were not usually thought of as being connected. . . .
>
> . . . Rosenberg . . . seems to think that I want economists to stand shoulder to shoulder with physicists, chemists and microbiologists against poets, metaphysicians, astrologers and Freudian analysts. Like McCloskey I do not wish to do this.[28]

These paragraphs record a shift in Weintraub's thought and work, one indicative of the influence that McCloskey has had

27. Readers interested in understanding the defects of Lakatos's methodology independent of the uses to which Weintraub puts it should read Hausman, *The Separate and Inexact Science of Economics,* section 9.8. Hausman not only identifies difficulties in Lakatos's approach but shows in detail why economic methodologists "need to adopt a greater measure of philosophical agnosticism." This is excellent advice, particularly if it is seen to include agnosticism about McCloskey's methodology manqué.

28. "Rosenberg's 'Lakatosian Consolations for Economists': Comment," *Economics and Philosophy* 3 (1987): 140.

over economists who at least initially turned to philosophy of science out of concern for the adequacy of their discipline. By the time Weintraub had written his next book,[29] the transformation had moved him completely from epistemology to rhetoric: "What began as an attempt to examine, and appraise, the work on stability of the competitive equilibrium has ended as something rather different" (p. 3). In the interim, Weintraub had discovered McCloskey, to whose "profoundly important book," *The Rhetoric of Economics*,[30] Weintraub credits his rethinking. Weintraub's rethinking has resulted in striking conclusions: for example, knowledge is constructed. This claim seems to mean that it is not knowledge at all (i.e., not justified true belief); rather, it is a sort of alibi negotiated among those whose activities will be held up for public scrutiny. And this goes both for economics and for the history and philosophy of economics: "If we do not want to argue any longer that economists attempt to test, or falsify, theoretical accounts by confronting those theories with data, if we do not believe that this framework provides us with a convincing account of scientific practice, we have a problem at the metalevel as well. . . . If we are not falsificationist in our conceptualization of economics, we are under no compulsion to remain falsificationist about our histories of economics" (p. 4).

What this passage means is that Weintraub has repudiated the notion not only that economics does or should proceed in accordance with some empiricist's conception of science, but that the study of economics as a discipline also needs to forgo the notion that there is some fact of the matter about how economics ought to proceed, some fact out there to be discovered by the philosophy of science. Instead, like the literary critic's approach to literature, we should treat economics as an "interpretive community." "I am left with a merely residual interest in philosophy. . . . I am interested in criticism, or appraisal in the general sense of general cultural criticism, with a focus on the

29. *Stabilizing Dynamics: Constructing Economic Knowledge* (Cambridge: Cambridge University Press, 1991). The quotations from Weintraub below are from the introduction to this work.

30. (Madison: University of Wisconsin Press, 1985).

texts or models or theories or evidence of economists instead of the canvasses of painters, or the films of directors or the experiments and lab records of biologists" (p. 8). But it cannot be a matter of indifference to economists whether their subject has the character of chemistry as opposed to literary criticism. Unlike literary criticism, matters of great public moment turn on the claims of economists: jobs, fortunes, even lives are at stake in the advice economists give to policymakers. To this extent what an economist says can make a much greater difference for a larger number of people than poetry, metaphysics, astrology, and psychoanalysis. As such, economics has a responsibility far greater than those disciplines whose only standard is that identified in rhetoric.

What is more important is that like Friedman and others, Weintraub's reading of the record of twentieth-century economic theory is one of steady increase in the predictive success of aggregate consequences, in what Lakatos would call its "empirical progressiveness." But if this were unarguably so, the entire question of appraising economics would be moot. There would be no need to show it satisfies some standard drawn up in the philosophy of science. At most its satisfaction of such a standard would tend to corroborate Lakatos's philosophy of science. It is the absence of agreement on empirical progressiveness that raises the methodological problems that Weintraub's book appears to address.

In his new book, Weintraub reports that reflecting on Weintraub's original claim, McCloskey commented that "accepting my [Weintraub's] reconstruction of a Lakatosian program, the consequences were simply that one now could model economic analysis in a philosophically coherent framework. But what was the payoff to this linkage? What more did we know about economics?" The question appears to have been rhetorical, and Weintraub's answer is not recorded. Like McCloskey I am inclined to say that we do not know much more about economics as a result of pressing it into a Lakatosian mold. But the questions about economics that we hoped to answer by so pressing it remain, and refusing to answer them will not make them go away. In the next chapter I attempt to answer at least some of them.

5

ECONOMICS AND
INTENTIONAL
PSYCHOLOGY

Economics is an inexact science, but there are many such disciplines. The puzzling thing about economics is that it seems no more exact than it ever was. Methodologists have traditionally compared economics to another inexact discipline, meteorology. One thing telling about the comparison is that in our life times, meteorology has made marked improvements in exactness.

The inexactness of disciplines is reduced as the precision of their explanations and the accuracy of their predictions increase. There are only two ways in which such improvements can be effected: either by more exact measurements of the initial or boundary conditions to which a discipline's theories are applied or by an increase in the precision of the theory's claims about the mechanisms that lead from initial conditions to consequences. In meteorology, presumably, the most important improvements have been made in the measurement, collection, and aggregation of data to provide the initial-condition inputs. In this chapter I argue that economic theory's character makes impossible both improvements in the specification of initial conditions and improvements in the generalizations of the theory itself.

Explaining or explaining away the persistently inexact character of the science of economics has been one of the stocks in trade of the philosophy of social science from the days of John Stuart Mill onward.[1] Mill and many followers have traced the in-

1. See Mill's *A System of Logic* (London: Longman's, 1949) and *The Collected Works of John Stuart Mill* (Toronto: University of Toronto Press, 1965–67),

112

exactness of economic theory to the presence of ceteris paribus clauses in the fundamental and the derived generalizations of economic theory. My explanation of the predictive limits of economic theory can be understood as the claim that such "other things being equal" clauses turn out to be ineliminable, and not even much reducible in scope. To show this we need to understand how ceteris paribus clauses work in general. Fortunately, Daniel Hausman has done this for us in *The Separate and Inexact Science of Economics*.[2] I shall adopt his account of the matter in order to explain my approach.

Consider a general statement of economic theory, either an axiom like "all agents' preferences are transitive" or a derived generalization like "whenever supply increases, price falls." Both of these propositions are insulated from falsehood by implicit ceteris paribus clauses. Their form is more like "ceteris paribus all F's are G's" or "ceteris paribus whenever c happens, e happens." Economists have long held that to be testable, economic general claims must eventually be purged of these ceteris paribus qualifications. To purge a ceteris paribus clause requires enumerating the other conditions which these clauses claim to be equal or held constant or cancel out. This will be possible only if the set of such conditions is finite, manageable to state, and measurable. As we identify the component conditions of this set, we reduce the inexactness of the relevant generalizations. But suppose there is no such finite manageable set of conditions that must be satisfied for the general statement to obtain. Suppose the conditions which must be equal are diverse, heterogeneous, unwieldy to combine, difficult to measure, and otherwise scientifically intractable. If there is no manageable set of conditions, there is in effect no scientifically interesting property that we could discover and substitute for the ceteris paribus clause in the generalization we are attempting to improve. Hausman's

---

vols. 2–4. For an excellent introduction to Mill's treatment of these topics see Daniel Hausman, "John Stuart Mill's Philosophy of Economics," *Philosophy of Science* 48 (1981): 363–85. Hausman's views about Mill's claims are expounded in *The Separate and Inexact Science of Economics* (Cambridge: Cambridge University Press, 1991).

2. Chap. 7.

approach to ceteris paribus clauses reflects this analysis (though he does not explicitly employ it):

> "*Ceteris paribus* everything that is an F is a G" is a true universal statement if and only if in the given context the *ceteris paribus* clause picks out a property—call it "C"—and everything that is both C and F is G. . . . In committing oneself to a law qualified with a *ceteris paribus* clause, one envisions that the imprecision of the predicate [C] one is picking out will diminish without limit as one's scientific knowledge increases.
>
> Thus to believe that, *ceteris paribus*, everybody's preferences are transitive is to believe that anything that satisfies the *ceteris paribus* condition and is a human being has transitive preferences. One need not be disturbed by intransitive preferences caused by, for example, changes in taste, because such counterexamples to the unqualified generalization lie outside [the extension of the predicate] C. . . . A sentence of the form "*ceteris paribus* everything that is an F is a G" is a law just in case the *ceteris paribus* clause determines a property C in the given context, and it is a law that everything that is C and F is also G. (Pp. 136–37)

The questions must emerge, When can we know whether there is in fact such a property as the ceteris paribus clause assumes, and why does a ceteris paribus generalization have explanatory power before we have identified this property? Hausman's answer to the second question is straightforward: "When one takes an inexact generalization to be an explanatory law, one supposes that the *ceteris paribus* clause picks out some predicate that, when added to the antecedent of the unqualified generalization, makes it an exact law" (p. 137). The first question, whether there is a property of the kind the ceteris paribus clause assumes, is the question of what justifies our present belief that there is an exact law behind our ceteris paribus generalization and that we will eventually discover the property which will convert the ceteris paribus generalization into the exact law. Hausman outlines three conditions ceteris paribus generalizations must satisfy for us to feel much confidence that they reflect as-yet-unknown exact laws: (1) reliability—"ceteris paribus all F's are G's" is reliable only if (perhaps making allowances for specific interferences) almost all F's are G's; (2) refinability—if we add specific qualifications, the generalization should become more reliable, or applicable to a wider domain, even though it continues to require a ceteris paribus clause; (3) excusability— when the inexact generalization fails, we need to be able to identify the specific interfering factor that prevented the generalization from holding.

Hausman's discussion is an important improvement on what is by now a generation of attempts to come to grips with the ceteris paribus character of economic generalizations.[3] In the light of this account of the nature and significance of ceteris paribus clauses, I want to explain what it is about the causal variables in economic generalizations that makes it impossible to satisfy the requirements of reliability, refinability, and excusability. Indeed, I hope to convince the reader that there is no property like C which when added to the antecedent of an inexact generalization of economic theory makes it an exact law. I do not argue that there can be no property (like C) that will convert inexact to exact laws, just that, as a matter of fact, there is none. Such an argument is in a broad sense a factual and not an a priori one. I argue that the burden of the evidence is against the existence of such a scientifically interesting property. My argument hinges on features of the economists' stipulations about individual rational choice—features philosophers and psychologists have uncovered but which economics has yet to assimilate. Another way to express the conclusion to which I come is again in terms of a notion of John Stuart Mill. Mill held that economics was a "separate" science, one in which the "major causes" of the phenomena of interest to economists can be identified ("the desire of wealth"), and that these major causes pretty fully explain the phenomena of interest. If my account of the ineliminable source of inexactness in economics is correct, then

3. See, in particular, Alexander Rosenberg, *Microeconomic Laws* (Pittsburgh: University of Pittsburgh Press, 1976), chap. 6, especially pp. 134–38; and Daniel Hausman, *Capital, Prices and Profits* (New York: Columbia University Press, 1981), pp. 120–39.

the presumption that economics is a separate science in Mill's sense must be surrendered; the causes are too multifarious for major ones to be identified.

It is easy to catalogue the limitations, idealizations, abstractions, and false assumptions of the theory of rational choice in microeconomics. Jon Elster has conveniently brought many of them together in "When Rationality Fails."[4] He divides the failures of rational choice theory into two kinds: failures of indeterminacy, when the theory does not issue in a unique prediction, and failures of inadequacy, when the prediction is falsified by events. He notes, rightly, that the second failure is more serious. Among the sources of indeterminacy in the theory's output, Elster notes the possibility of multiple optima. More serious problems for the theory are generated by the incompleteness of preference orders, especially between alternatives difficult to compare, like career choices, for example. Problems emerge from uncertainty about available alternatives and of course from strategic interaction among rational agents. And finally there is the iterated rational choice problem of how much should be invested in acquiring information that will be used to make rational choices.

The theory of rational choice is inadequate, that is, makes false predictions, because people are irrational. Elster enumerates the ways in which irrationality emerges: weakness of will, when agents choose the alternative less preferred, or seem to do so; wishful thinking and its opposite—phenomena like those of cognitive dissonance, in which desires are misrepresented to oneself so that they correspond with beliefs; and mistakes about probabilities and their calculations.

One response to all these problems is to hope that they will cancel one another out in the aggregation of individual choice into market phenomena. The argument behind this response begins by helping itself to Herbert Simon's distinction between procedural and substantive rationality.[5] Procedural rationality

4. In *Solomonic Judgements* (Cambridge: Cambridge University Press, 1988), chap. 1.
5. "From Substantive to Procedural Rationality," in *Method and Appraisal in Economics,* edited by S. Latsis (Cambridge: Cambridge University Press, 1976).

characterizes the psychological process through which rational choices are arrived at. Substantive rationality characterizes those actual choices that satisfy criteria of constrained preference maximization. Economists conventionally argue that they can leave issues of whether choice is procedurally rational to psychology, since all they need is the assumption that it is substantively rational. And of course if neoclassical theory's predictions about aggregate phenomena were good enough and/or showed the kind of improvement we expect of a scientific theory, then this would be a reasonable view to take. One could hold that by and large in the aggregate, individuals are substantively rational and we may calmly write off the problems in its assumptions as random noise that can be ignored by economists primarily interested in the market. But of course neoclassical economic theory is not so fortunate. Its predictive powers have been weak enough for long enough that we need to seek an explanation for the weakness. It is natural to seek its problems in the imperfection of its assumptions about individual behavior and to seek their solution in the improvement of these assumptions.

Of course, it is no wonder that a series of assumptions about individuals that can go wrong in so many ways should when aggregated fail to produce improvable conclusions about large-scale economic processes. Were economics a normal discipline, one would expect that at least a fair amount of its intellectual capital would be invested in attempts to reduce the indeterminacy and inadequacy of the theory of rational choice by improving on the realism of its assumptions. But no such investment has been made. Economics as a discipline has made little effort to improve what Hausman calls the reliability, refinability, and excusability of its ceteris paribus generalizations. Economics has in fact responded to problems of indeterminacy and inadequacy, not by improving the strength of its claims about individual choice, but rather by weakening them to the point of vacuity.[6]

6. See the discussion of this trend in Alexander Rosenberg, "A Skeptical History of Economic Theory," *Theory and Decision* 12 (1980): 79–93; and idem, *Sociobiology and Preemption of Social Science* (Baltimore: Johns Hopkins University Press, 1981), chap. 4. Several authors in R. Hogarth and M. Reder,

EXPECTATIONS AND PREFERENCES, BELIEFS AND DESIRES

It is easy to show that the fundamental explanatory strategy of economic theory is of a piece with that of our ordinary explanations of human action. It is also easy to show that economists' interpretations of this strategy that seek to sever its connection with "folk psychology" are unavailing. Some of these interpretations involve Friedman-like special pleading about the intended domain of economic theory; others involve tendentious reinterpretation of the theory of economic choice. Having discussed these approaches elsewhere,[7] I will only sketch them briefly here and otherwise focus on the light that recent work in philosophy of psychology can shed on economic theory.

Economics proceeds by formalizing commonsense explanations of action into a theory of rational choice. Consider the characteristic features of ordinal utility theory:

1. Comparability: For all possible pairs of commodities, the rational agent prefers one to the other or is strictly indifferent between them.

2. Transitivity: For any three commodities, a, b, c, all rational agents who prefer a to b and b to c prefer a to c.

3. Rational agents maximize: They choose the combination of commodities within their budget constraints that they prefer most.

4. Economic agents are rational: They act in accordance with 1–3.

Preference seems unarguably a matter of strength of desire. In the marginalists' interpretation of cardinal utility, some amount of psychic pleasure, satisfaction, utility, or the like is associated with each commodity. Though ordinal utility surrenders the units, it does not surrender the attitude. Similarly, the budget constraint constrains via the agents' knowledge of their endow-

---

eds., *Rational Choice* (Chicago: University of Chicago Press, 1986), raise and respond to the same problems. See especially Herbert Simon, "Rationality in Psychology and Economics," pp. 25–40.

7. *Microeconomic Laws* and "A Skeptical History of Economic Theory."

ments and the prices of the commodities for which they have a preference order. It is of course logically possible that the size of endowments and the prices of commodities interact directly with preferences, without somehow being represented in the agents' minds. But surely no one would appeal to this bare possibility in order to argue that economics has nothing to do with psychology. And of course economic theory does not do so. In fact, the standard boundary condition on the presentation of rational choice theory is that individuals have complete information and that they operate under conditions of certainty. The assumption of complete information would be gratuitous if the theory was intended to be neutral on how budgets constrain and prices affect choice. When the theory is extended to conditions of risk, the role of beliefs in choice becomes even more manifest, as we shall see.

It is assumption 3 that makes the affinities of rational choice theory and folk psychology most manifest. Our ordinary explanations of action involve showing that the action an agent actually performs is the one he or she believes will lead to some goal or other he or she actually embraces. This explanatory strategy works because folk psychology endorses an (often unstated) principle of human action to the effect that individuals always choose those actions which they believe to be most suited to the attainment of their goals. Sometimes this principle is held to be a definition of action, sometimes a contingent generalization. But the claim that economic agents act so as to attain their most preferred available alternative is pretty clearly a straightforward variant on this folk-psychological principle.

Economists have long been uncomfortable with this fact about the way that economics explains, and they have sought interpretations of the theory of rational choice that sever its connections with folk psychology. Such interpretations usually proceed by attempting to deny any economic commitment to psychologically represented preferences. Revealed preference theory tells us that so long as economic agents behave in accordance with assumption 2 we can derive all the economic theory we need for explaining the aggregate consequences of individual choice. And since economics has nothing to say about the causes of choice, owing to the exogenousness of preference, we

119

should interpret ordinal utility theory as a behaviorist would, instead of treating preference as causing behavior and as "revealed" by it. We should redefine "X prefers A to B" to mean "X actually chooses A when both A and B are available." Then by offering the agent a wide variety of paired choices, we can construct a map of those the agent is indifferent between, and thus we are on our way to supply and demand curves without even talking about the agent's desires. Revealed preference thus turns out to be a misnomer. For in this theory agents have no preferences to reveal by their choices. They simply make choices. We draw up an indifference map on the basis of them and employ it to predict their choices. But we do not attribute any psychological reality to the indifference map; we do not treat it as reflecting or revealing agents' otherwise hidden desires.

Revealed preference theory in effect locates the starting point of economics in actual behavior, in the consequences of our decisions. It surrenders all claims to the explanation of choices by individuals, leaving this matter to psychology. In this interpretation, economic theory cannot be taxed with failure to provide a predictively improvable theory of individual choice behavior, because it takes behavior as its explanans not as its explanandum. The theory of rational choice is just a useful device for systematizing such behavior, one we would be mistaken to interpret literally.

Of course unilaterally changing the subject of economics by redefining its explanatory variables has little to recommend itself as a solution to the problem of why economic theory cannot explain individual behavior very well. After all, defining the problem of individual choice out of existence as a problem for economics does not make it go away. It just shuffles off the responsibility for the problem to some other discipline. Moreover, revealed preference is an indefensible interpretation of rational choice for independent reasons. And it will not be enough to sever the connections between the theory of rational choice and folk psychology in any case. The theory is indefensible because it does not do what it claims: embracing it, economists are still committed to the representation of preferences in the heads of economic agents. And even if preferences were not in the head, the role of beliefs, expectations, and constrained maximization would preserve the affinities of rational choice theory to folk psychology.

Under conventional specification of available commodities economic agents violate the transitivity requirement of revealed preference theory with monotonous regularity. Having expressed a preference for caffeinated coffee over milk at breakfast, they choose milk over decaffeinated coffee at lunch, and the latter over caffeinated coffee at dinner. Well, it is pretty obvious that this is no sign of irrationality and that the commodities need to be dated and perhaps also bundled with other commodities before we explore individual preference transitivity. But sometimes people's tastes seem to change. Having preferred caffeinated coffee to milk on Monday morning, they prefer decaffeinated coffee to either on Tuesday, and milk to either on Wednesday, even though so far as investigation reveals other things really are equal. Of course, the reasonable thing to say in this case is that the individual's tastes change over time. Perhaps they reflect some sort of "hysteresis." Caffeinated coffee on Monday increases the preference for decaf on Tuesday, and it in turn makes the individual crave milk on Wednesday. Tastes change. But the theory of revealed preference cannot distinguish between irrationality—violations of its transitivity requirement—and changes in taste. Why? Because tastes are just the sort of things revealed preference theory has no place for. The whole point of the theory is its silence about what agents like, as opposed to what they choose.

If commodities are finely enough differentiated or bundled together in just the right way, then of course there might not be any such thing as change in taste. Take our coffee, decaf, milk case. If the arguments of the utility function are actually coffee-after-milk, decaf-after-coffee, milk-after-decaf, then the violation of transitivity disappears, as well as the change in taste. Economic theory can fob off the complaint that these commodities are ad hoc constructs designed to avoid the problem of changes in taste versus intransitivity, for it already must hypothesize a vast diversity of dated and located commodities that ordinary commerce does not recognize in order to prove the existence of a market-clearing general equilibrium. So why not help itself to them at an earlier stage in the theory's development?

However, now an interpretation that began with the motivation of making the theory of rational choice more testable by giving it a behaviorist interpretation has had the opposite result: given the apparently gerrymandered commodities among which the agent "reveals" preferences, it is impossible ever to detect a violation of transitivity. The agent never faces two choices in which the same commodity figures both times. The agent never has a chance to prefer a to b at time t_1 and b to c at time t_2 and c to a at time t_3. b-in-the-presence-of-a-at-time-t_1 is not the same commodity as b-in-the-presence-of-c-at-time-t_2.

But even if revealed preference theory had enabled economists to banish preference from their theory, it would have done nothing about expectations. And taking expectations seriously is impossible unless we take preferences seriously as well. The inextricable connection between them is the lesson of Von Neumann and Morgenstern.

The theory of expected utility provides both a means of measuring the strength of preferences (as opposed to merely ordering them) and at the same time a way of dealing with uncertainty, a means of dispensing with the most idealizing of economic assumptions, the assumption of perfect information. Beginning with ordinal preference $a > b > c$, we measure the strength of preference by finding that lottery ticket offering a certain probability, p, of getting a against $1 - p$ of receiving c, such that the agent is indifferent between it and the certainty of receiving commodity b. The greater the strength of the preference for a over b, the greater the risk the rational agent will be willing to take in lottery tickets offering a chance of a against a chance of c. And this risk is quantitatively measurable in probability values of receiving a or c, so that these numbers enable us to attach a number to the strength of preference of a over b, and not just its order. But notice, this method of attaching weights to preferences requires two things: first the agent must have beliefs about the probabilities of outcomes (i.e., expectations), and second, he or she must make choices among lottery tickets. Holding the beliefs about probabilities constant and observing the lottery choices, we work backward to the preference structure.

The Von Neumann–Morgenstern recipe for quantifying preference is also one for dealing with imperfect information. If we
already know the agent's strength of preference among alternative outcomes, then by offering him lottery tickets between these
alternatives, at varying probabilities, we can determine the probabilities he attaches to these outcomes.

The upshot is that the Von Neumann–Morgenstern theory
of expected utility provides a means of calculating the strength
of an agent's belief about the occurrence of any alternative in
which he has an interest; and it provides a means of determining the strength of his preferences among these alternatives.
But it cannot do both of these things "at the same time" so to
speak. In order to determine strength of belief, we must observe
behavior, holding strength of preference constant. In order to
calculate strength of preference, we must observe behavior,
holding strength of belief constant. In both cases, we may read
some of the agent's psychological states from behavior, but only
we already know the agent's other psychological states.

It will be important for our argument to note that the Von
Neumann–Morgenstern approach does not provide independent methods of "holding constant" preferences or beliefs. In
order to infer beliefs from choices, we must use the theory of
expected utility to calculate preferences beforehand. In order to
determine preferences from choices, we need to use the theory
of expected utility to calculate the strength of beliefs beforehand.
So there is no way to test the hypothesis that individuals are
expectations-constrained, expected-utility maximizers, because
we need the hypothesis to determine the expected utilities the
hypothesis predicts they will maximize and to determine the expectations that constrain their maximization.

Now, of course, the traditional economist's attitude toward
preferences and expectations—desires and beliefs—is to treat
the former as exogenous and the latter as perfect, at least at
the outset. Preferences are left to psychology for elucidation
and imperfect expectations to further improvements and am-
plifications on the pure theory. In fact the theory of rational
choice under certainty is just a limiting case of Von Neumann–
Morgenstern expected utility, where agents unerringly attach

probabilities 1 or 0 to every possible outcome, so that we can offer them no lotteries and thus cannot construct a cardinal utility structure for their preferences.

Treating tastes and preferences as exogenous does of course reduce the exposure to test of the theory's predictions about individual choice. We can never tell whether an apparent violation of transitivity is the result of a theory-falsifying act of irrationality or a theory-neutral change in taste. Similarly, our theory's development under the assumption of perfect information reduces its applicability to cases of imperfect information, that is, to the "real world." However, one must walk before one can run, and the assumption of perfect information is one we have to make if we are even to get started.

Suppose, however, that we can show there is in principle no way ever to measure the strength of preference or belief independent of one another and of the hypothesis that agents are utility maximizers? Then, the theory is condemned to perpetual predictive weakness. It will never run any faster than it does already.

THE PROBLEM OF IMPROVABILITY

If we are to apply, test, and improve the explanations we make with the hypothesis that agents engage in rational choice, we need to measure the "initial conditions" to which we apply this hypothesis. Especially if we want to improve our predictions, we need to improve our measurements of the states of the agent to which we apply the theory in order to secure predictions about behavior.

To see this, compare the ideal gas law:

$$PV = rT.$$

To apply this law to predict the volume of a gas, one needs to measure the gas's temperature and its pressure, and of course one needs to measure the volume in order to assess the accuracy of the prediction. A thermometer and a pressure gage, as well as a way of measuring volume, are needed. Given readings on the thermometer and the pressure gage, we can plug them into the equation, and if we know the value of r, we can calculate the volume of the gas, which is then compared with the actual volume. If the predicted value differs very much from the observed value, we can surrender the ideal gas law, or we can call the accuracy of the thermometer and pressure gage into question. If our prediction is badly off, we can even call into question the theories that guide construction and calibration of such instruments, though doing so looks suspiciously ad hoc.

Suppose our prediction agrees with the observed value of volume to a couple of decimal places, but diverges thereafter. Assuming we trust our instruments, how can we improve our prediction? There are two ways: we can try to improve the accuracy of these instruments, or if we think them accurate enough, we can attempt to improve our gas law. In the history of chemistry and thermodynamics, both of these tacks have been taken. Thermometers have been improved, and improvements have been made in the ideal gas law, adding variables and coefficients that improve its realism. In fact, improvements in the gas law were contingent on improvements in instrumentation, because it was only with improved gages that we were able to measure the temperature and pressure of gases at extreme values, and it is at these values that the ideal gas law breaks down and calls for improvement. Once enough was learned about the behavior of gases, of course, the tables were turned, and the successors to the ideal gas law were called upon to help design means of measuring temperature that exploited its relations to the pressure and volume of a gas.

But suppose that the only way to measure the temperature of a gas is to measure its volume and pressure and then plug these values into $PV = rT$. That is, suppose that the only sorts of thermometers available depended for their accuracy on the truth of the ideal gas law. If this were the case, then the gas law would be vacuously true and useless in predicting the behavior of gases. No set of observed values of pressure, temperature, and volume could disconfirm the law, because it is used to determine these values. That makes it vacuous. It would be predictively useless, because to predict the pressure of a gas, we would need to know its temperature, and the only way to acquire this knowledge is to determine its pressure, which is what we want to predict.

Evidently, it is a requirement for the predictive power of a theory that the instruments we employ to predictively apply it to initial conditions be independent of the theory to be applied.

Now, what we need to apply, test, and improve our theory of rational choice is the equivalent of a thermometer: something that will enable us to measure the initial conditions to which we apply rational choice theory, that is, the preferences and expectations of agents. And this "thermometer"—this means of telling what agents believe and what they want—must be independent of the hypothesis of rational choice: that is, the reliability of the instrument we use to measure strength of desire and degree of belief must not hinge on the truth of the theory of rational choice.

However, this is just what we cannot get for the theory of rational choice, because of the nature of desires, beliefs, and actions. There is no way to tell what a person believes unless we already know what he wants and how he acts; no way to tell what a person wants unless we know what he believes and how he acts; no way to tell what a person will do unless we know what he wants and believes. The only way any two of these three factors can lead us to a prediction about the third is via the theory of rational choice.

The easiest way to see this is to consider the problems of a behavioral approach to mental states. The idea of philosophical behaviorism was initially to show that statements that purport to refer to unobservable mental states are really disguised claims about behavior. Or at least we can substitute statements about behavior for statements about the mind, without loss of cognitive content.

So, the statement that Jones believes it will rain today is to be translated into or defined in terms of Jones's disposition to take an umbrella. But Jones will take one only if he wants to stay dry, believes an umbrella will keep him dry, doesn't mind others thinking him old-fashioned for carrying one, and believes an umbrella is available for his use. Of course all these conditions and the indefinitely many more like them may be absorbed into a ceteris paribus—other things equal—clause of the sort economists are familiar with. The trouble is that these clauses themselves make mention of the very mental states a behavioral definition sets out to avoid. So, any acceptable definition of a belief in terms of the behavior it leads to will involve specifying the agent's desires and other beliefs and holding both constant:

Agent x believes that action a is the best means to goal g if and only if he does action a and has goal g.

The same applies to behavioral definitions of desires:

Agent x wants g if and only if he does action a and believes that a is the best available means of securing g.

And when is a movement of the body an action, as opposed to a sneeze, a shudder, a blink? Only when the behavior is caused by the agent's beliefs and desires:

Behavior b is an action only if the agent has goal g, believes that behavior b is the best available means to secure g, and the belief and desire cause behavior b.

One might suppose that we need not wait for agents' behavior to reveal their beliefs and desires. Sometimes we can just ask what they believe and what they desire. The method of just asking may not work every time, but surely it will work some of the time. The trouble is that the method of asking, for all its merits, is just an instance of the behaviorist's technique of determining desires and beliefs: it works only by helping itself to just the assumptions about the mind behaviorism is supposed to forgo.

Suppose we ask someone whether he believes it will rain today. Our interlocutor produces the noise "yes." First we need to consider whether the production of this noise was speech or a peculiar sneeze. That is, we must decide whether it was caused by some combination of desires and beliefs or by events in the medulla. Assuming the first, we now need to decide on which combination of desires and beliefs led to the action of saying "yes." The trouble is that there are an infinite number of different combinations of desires and beliefs that can lead to an affirmative response to the question, Do you believe it will rain? For example, the desire for a cigarette combined with the belief that saying "yes" to the next question one is asked will lead to one's securing a cigarette will cause an agent to say "yes" to our question. Similarly, the belief that "yes" is the way to signal dissent in English coupled with the desire to deceive the questioner about one's meteorological beliefs will lead to the answer "yes." Of course these combinations are farfetched. But that is the point. There are an infinite number of such combinations of belief and desire that together will produce the action we expect. For this reason, the action of saying "yes" in answer to our question is just another example of a behavioral approach to mental states and an illustration of its failure to avoid assumptions about the mind. Moreover, even when our interlocutor wishes to signal assent to our question and believes that saying "yes" is one way to do so, he or she may say "yup" or "ya" or "u-huh" or nod the head or express assent in any of a number of other ways. Working back from one or another of these behaviors to what the interlocutor believes and wants requires further assumptions about intentional states. Verbal responses to direct questions provide perhaps the most accurate basis from which to infer back to intentional states, but we can acquire these only under the most artificial experimental conditions. Inference to mental states from every other sort of behavior is fraught with even graver doubt and ambiguity.

What does all this mean for the theory of rational choice? Well, if some sort of behavioral test is the only way of establishing the initial conditions to which we apply the theory of rational choice in the prediction of behavior, then there can be no independent measure of the strength of desires or degrees of beliefs about alternatives. The theory of rational choice is condemned to predictive sterility.

But what if there is some other means of determining what people believe and want, one that does not rely on the theory of rational choice. Surely there is an alternative, at least in principle. After all, desires and beliefs are states of an agent's brain that cause behavior. It must therefore be possible to somehow read an agent's mental states from his brain states. Although reading beliefs from the brain will be of little practical help to rational choice theory (after all, we cannot consult a neuroscientist every time we wish to predict the behavior of an individual consumer, let alone a market of them), it is of the first philosophical importance. It means that there is at least in principle a means of telling what people believe and desire independent of their actual choices. So, explanations employing the theory of rational choice are in principle capable of improvement, as we require of a scientific theory. The situation is rather like what would obtain in thermodynamics if the only way of measuring a gas's temperature independent of the gas law was fiercely difficult but theoretically possible. In that case, improvements in the predictive accuracy of the ideal gas law would be hard to come by, but at least possible.

On this analogy, the limitations on the predictive accuracy of the theory of rational choice, both for individual actions and for the aggregation of them, rest on the coarseness of our measure of the beliefs and desires of agents. If we only had better measuring devices and the computers to aggregate them, we could consistently improve our predictions. Given the complex character of our own neurology, there is no better practicable measuring device for beliefs and desires than the theory of rational choice itself.

It would be pointless, however, to appeal to the eventual prospects of neuroscience to provide access to our beliefs and desires independent of behavior as any part of the justification for pursuing intentional explanations of behavior. The more we learn about the way the brain stores information of the sort to be represented in our beliefs and desires, the more it becomes apparent that there is no fact of the matter about exactly what we want and exactly what we believe. That is, the propositional content of these states is misleadingly precise in the preferences and expectations it attributes to us.

The upshot of the intentional character of the explanatory variables of economic theory is obvious. We cannot expect the theory's predictions and explanations of the choices of individuals to exceed the precision and accuracy of the commonsense explanations and predictions with which we have all been familiar since prehistory. And if our predictions of the behavior of individuals faced with individual choices are fated to be at best vague and imprecise, what can we expect when we aggregate individual behavior? It is improbable that we can improve on the accuracy of claims about the aggregation of human choices without improvements in our accuracy about individual choice. As the history of science suggests and, for that matter, as the rational expectations economist rightly holds, improvements in microfoundations are the best ways to improve the accuracy of macropredictions.

Compare the kinetic theory of gases. Though it was well known that the ideal gas law fails for certain gases beyond certain limits of temperature, volume, and pressure, there was no hint of how the formula could be corrected, except by adding variables that reflected the elimination of idealizations in the assumptions of the kinetic theory of gases. Starting with the false assumption that molecules are infinitely elastic, we can improve on the gas law by incorporating a term that reflects their finite elasticity. Then given the false assumption that molecules are point-masses, we can incorporate a term in the gas law that reflects the actual volume molecules take up. The result is a more accurate gas law, whose construction is guided by improvements in the assumptions of the kinetic theory about individual gas molecules. Without such guidance, the best we could have hoped for was a more accurate equation for each different gas derived by curve fitting as more and more data disconfirming the ideal gas law came in. But such curve-fitted formulae would lack real explanatory power, fail to reveal what gases had in common that unified their behavior, and be at the mercy of our measuring instruments.

The aggregate generalizations of neoclassical theory are in much the same position as the ideal gas law before the discovery of independent means of determining the intermolecular forces and the incompressibility of gas molecules. However, given the intractable problems of intentionality recorded above, there is no hope of parallel discoveries about preferences and desires that will actually influence the shape of demand or supply curves in the way that discoveries about molecules influenced the shape of curves relating pressure, temperature, and volume.

Reflections like these should help us understand the indifference with which economists have met recent work on rational choice theory by psychologists. Psychologists and even some economists have, over the last decade and more, subjected rational choice theory to a variety of experimental and observational tests. By and large these tests have revealed forcefully the limits on rational choice theory's assumptions about individual agents, both each assumption taken alone and the package of them considered together.[8] But there is no way systematically to incorporate the forces that account for the findings of these psychologists into the model of choice as constrained-preference maximization except by treating them as further expectations and additional preferences, whose effects cannot be measured without assuming the substantive rationality of agents. Accordingly, economists will be unable to see what the fuss is all about, for their theory remains unchallenged from the direction of psychology, despite the theory's predictive weakness and its likelihood of remaining so.

ECONOMICS AND LINGUISTICS: NELSON'S GAMBIT

Alan Nelson has challenged this gloomy picture of the prospects for economics as a predictively powerful and explanatorily precise science of human behavior and its aggregation.[9] Nelson begins, rightly, by taking economics in general and the theory of rational choice in particular at face value. Neoclassical economic theory is undoubtedly concerned with the explanation and prediction of individual economic behavior, and reinterpretations of its formalism to circumvent the problems facing its explanation and prediction of such behavior are unavailing. Revealed preferences notwithstanding, the theory is committed to the psychology of preferences and expectations. Nelson wants to "consider the consequences of basing microeconomic theory" on a claim like

(5) Utility functions are psychologically real in the sense that, (i) given appropriate idealizations, making a rational economic choice requires the agent's psychologically accessing the utility function as well as the economically

8. For a convenient summary of this research directed specifically at the issue under discussion here, see A. Tversky and D. Kahneman, "Rational Choice and the Framing of Decisions," and H. Einhorn and R. Hogarth, "Decision Making under Ambiguity," both in *Rational Choice,* edited by R. Hogarth and M. Reder, pp. 67–94, and 41–66.

9. "New Individualist Foundations for Economics," *Nous* 20 (1986): 469–90. Page references to Nelson in the remainder of this chapter are to this work.

relevant information concerning prices, the budget constraint, etc.; (ii) the utility function is then maximized subject to the constraint set by this information; (iii) the agent's choice is determined by the result of this process. (P. 476)

Nelson notes that claim 5 does not require agents to undertake these calculations consciously, nor need any part of the process be introspectively accessible. Indeed, there is no reason why economic choice should be different from routine linguistic tasks that require a good deal of processing, none of which is conscious or introspectable. The analogy between choice and speech and a parallel analogy between neoclassical theory and linguistic theory will prove to be a telling part of Nelson's argument. Relatedly, Nelson points out that the procedures whereby an agent maximizes utility need have little to do with the mathematical procedures economists use to effect their computations.

In this interpretation, it becomes appealing to characterize a special research program, which Nelson dubs "psychoeconomics," whose aim is to specify empirically mathematical forms that the utility functions take, thereby giving empirical content to the theory as a whole:

The psychoeconomist will hypothesize a functional form for an individual's utility function. In the beginning this will no doubt be a simple first approximation which is easy to work with. The investigator is likely to include only a few arguments in the function and assume that any other arguments will have effects that are small enough to be ignored in a first approximation. For example, something mathematically convenient (like this Cobb-Douglas utility function) may be chosen for an individual economic agent, i:

$$(6) \ U_i = a_i \log z_1 + (1 - a_i) \log z_2$$

where U_i is the amount of utility, a_i is a constant, z_1 is the amount of food consumed, and z_2 is the amount of shelter consumed. Then using ordinary econometric techniques, the scientist can estimate the value of a_i. The hypothesized function (6) can then be tested in routine fashion against data for real economic agents and actual values of z_1 and z_2. In the likely eventuality of poor agreement between the predictions of (6) and the observations, our scientist can introduce auxiliary hypotheses to account for the discrepancy or (6) can be discarded. . . . the parameters of (6) can be altered, or the z's appearing in (6) can be modified . . . , or the functional form of (6) can be changed, etc. Just about anything can turn out to be helpful in guiding the formation of hypotheses. . . . After this procedure is repeated many times we would expect to come up with utility functions which did an acceptable job of covering the data for given individuals and of predicting the results of future experiments on these individuals. If these expectations were not fulfilled it would be prudent to suspect that something was radically wrong with the whole approach. We might even become convinced that postulating utility functions was not a very good start in the construction of a scientific theory of economic choice. (P. 477)

Once we have framed utility functions for enough individual agents, we can hope to find generalizations about them: "For instance, it would be very interesting to discover that all utility functions were Cobb-Douglas in seven important commodities. . . . The more of these generalizations there are, the more powerful the explanations provided by the theory will be" (p. 478).

Certainly standard empirical methods bid us proceed the way Nelson suggests, and it sounds easy. But as experimental economists will report, undertaking the construction of a real utility function for a real person by observation of his or her actual choices either in an experimental setting or in "nature" is extremely difficult. In the real world there are too many factors to control, and in the laboratory, it is difficult and expensive to provide meaningful choice situations that will provide a utility function of use in predicting behavior outside the laboratory. Moreover, the intentional character of choice and preference makes it impossible to improve function (6) beyond a certain point, and that point is well beneath what we require for improvements in the predictive accuracy of common sense. To illustrate, suppose that (6) is initially not in accord with the data. Accordingly, we consider adding auxiliary hypotheses or altering and adding to the z's of (6). Note, however, that any auxiliary hypotheses will be about the agent's other preferences or the agent's expectations about z_1 and z_2, food and shelter: it is only via the agent's desires or beliefs that other variables will influence his actions. In order to calibrate his preferences and expectations we are going to need another utility function, of at least the same complexity as (6). How shall we set about acquiring this one? A regress is in the offing, as the preceding section would lead us to expect.

An attempt to add to the variables of (6) is vexed by the same problem. Doubtless, for a finite body of data about choices, we can fit a function like (6). The trouble is that as new data come in, the best we can hope for is post hoc corrections in (6) but no general account about whether and how utility functions change as more choices, and more information, become available, since we would need such an account to identify those choices and measure the incoming information to begin with.

All this does not mean that the enterprise of psychoeconomics is misconceived; it means only that it is doomed to failure. Given traditional conceptions of science, its failure augers ill for microeconomics, the discipline which presumes that utility functions like (6) are in principle available. So, Nelson asks why traditional economists find the research program of psychoeconomics wholly without interest. For him this is an important question, because he believes that psychoeconomics is both methodologically necessary for the advancement of economic theory and a potentially doable enterprise with worthwhile scientific results for the explanation and prediction of behavior. For our purposes the question is equally important, because answering it sheds light on the aims of economists and the nature of their theory.

Given the evident possibility of the line of inquiry Nelson calls psychoeconomics, it is puzzling to him that the ruling orthodoxy in economics rules out any contribution from psychology. He wonders what it is about this line of inquiry that economists find so repellent. The obvious first and discreditable reason Nelson entertains is that this sort of research exposes microeconomic theory to disconfirmation. If the curves eventually discovered violate transitivity or concavity or continuity requirements on the mathematical shape of utility functions, as we might well expect, there would seem to be little alternative to surrendering neoclassical theory. But, claims Nelson, this pessimistic conclusion is premature. There is a way of reconciling neoclassical theory with the empirical disconfirmation of its assumptions about the intentional causes of economic choice. The reconciliation exploits an analogy with similar circumstances in theoretical interaction between psychology and linguistics.

The aims of contemporary theories of transformational linguistics are twofold: (1) they should generate all and only the grammatically well-formed sentences of a language (presumably by derivation in an axiomatic system), and (2) they should explain linguistic data, like ambiguity, for instance, or perceived similarity, complexity, etc., by assigning structural descriptions to each sentence. Typically these theories proceed by postulating phrases and operations over these phrases that will construct sentences, convert statements into questions or commands, determine emphasis, and generate other linguistic items. Of course to explain linguistic behavior, this machinery must be assumed to be represented in people's brains and to be causally responsible for aspects of their speech.

The trouble is that actual speech behavior badly disconfirms these transformational theories of grammar: "Most humans produce a very substantial number of utterances that cannot be generated by the grammar, i.e., ungrammatical sentences. In fact there are undoubtedly a great many linguistic agents (perhaps it is most of them) who produce mostly ungrammatical speech. . . . On the other hand, it is well known that there are many classes of sentences that are grammatical, but are, nevertheless, virtually impossible for any human to understand under normal circumstances" (p. 480). For instance, suppose someone says to you, "It's not true that I couldn't fail not to disagree with you less." Quickly, decide whether this person assented to your views or not. If you can decode this sentence correctly, then it is easy to add further negations until you cannot.

The problem of accounting for actual verbalizations appears

to be quite similar to the one facing economics: pure theories largely unable to account for most of the data, the actual behavior of agents or speakers. Linguistics deals with this problem by drawing a distinction between the linguistic *competence* of agents and the linguistic *performance* of agents. Our linguistic competence includes two striking abilities: the ability to code and decode a potentially infinite number of sentences and the ability to classify sentences as grammatical or not with only a trivial number of exceptions. This theory of speaker/hearer competence appeals to "properties of a particular psychological entity, a grammar. . . . which is said to be internalized by the agent" (p. 481). A theory of actual *performance* explains behavior (as opposed to capacities) as resulting from the "interaction of the agent's competence with mechanisms postulated by other psychological theories that deal with how competence is used" (p. 481). Where theories of these two types develop, scientists may expect to be able to investigate the capacities of a system, without anxiety about how disconfirming data can be accommodated to the capacities. Dealing with disconfirming data is a project for the performance theorist, who deals with interfering and limiting factors that degrade the perfect exercise of competence. Of course, in the case of something like language, the performance theory cannot be expected to include every factor that interferes with perfect exercise of competence, because there are indefinitely many such factors. There must be a relatively limited number of interfering factors that, together with other causes and competence, actually generate a significant portion of the data. As Nelson points out, whether the distinction between competence and performance is useful in psychology or elsewhere "depends on how the facts turn out." He adds, "it is probably most realistic to expect these kinds of [competence and performance] theories to develop simultaneously with advances on one front feeding back into the other" (p. 481).

With the performance/competence distinction in hand, Nelson turns back to microeconomic theory:

Economists can study the implications of having certain kinds of utility functions in different economic environments working either completely abstractly or with actual utility functions determined with the aid of psychoeconomists. Moreover, they need not fear its turning out to be the case that any of the functions so obtained are empirically false, or even just inaccurate generalizations of the real functions. Discrepancies between the *competence* of economic agents based upon their ideal ability to maximize the utility functions which we now assume they really do possess, and the actual *performance* of these agents in the marketplace (or even in the laboratory) can be accounted for in terms of interfering factors which are irrelevant to abstract microeconomic theory. It can be hypothesized that this interference will eventually be satisfactorily explained as resulting from things like misinformation, imperfect memory, limited calculating abilities, etc. (P. 483)

Thus, the theory of *Homo economicus* is no different from the linguistic theory of the *ideal speaker*. Both are partial theories, because they do not make predictions until coupled with "some psychologically determined specifications of the functional relationships that [they] contain." The important point is that the prospects of performance theories, in both linguistics and economics, make it permissible to establish and extend pure theories of competence, despite their lack of independent payoff in predictions. Thus, the basic assumptions of microeconomic theory "can be well-insulated from disconfirmation without completely closing all the windows of vulnerability." So, Nelson notes, the view I defended in 1983[10] can at least be explained, if not vindicated:

> *In its present form,* microeconomics does not *look* like an empirical science partly concerned with individual behavior. Its lack of a genuine psychoeconomic treatment of individuals makes it look more like a branch of mathematics. This defect . . . can be removed through the practice of psychoeconomics. The relation between psychoeconomics and the ordinary theory is at least as sound in principle as the relationship between psycholinguistics and linguistics. Fur-

10. "If Economics Isn't Science, What Is It?" *Philosophical Forum* 14 (1983): 296–314.

thermore, proceeding in this way would make it possible to preserve the considerable achievements of micro-economics, quite unlike the suggestions made by many other critics of the theory. (P. 487)

In assessing this treatment of microeconomic theory, perhaps the first thing to note is that it is utterly at variance with the actual thrust of interpretation in microeconomics. Economists since Samuelson have strained to prune the theory of rational choice of its psychological content, to the extent of adopting revealed preference theory, according to which nothing need be supposed about what is going on in the head of the agent. The result is an interpretation that maximizes the theory's chance of disconfirmation; yet economists continue to cling to it. But here we have an interpretation that maximizes the theory's psychological reality and yet protects the theory from premature empirical test by treating it as an account of competence for which no performance theory has yet been offered. This interpretation is unlikely to attract the interest of economists, for the interpretation will oblige them to undertake the provision of a theory of performance, in which economists have no interest. The second reason is that if a theory of performance bulks very large in the story of how competence and behavior are really linked, then agents' actual behavior will only rarely be rational. Interfering factors will operate almost invariably, and almost none of the interesting aggregate results which economic theory includes, and which are supposed to be explained by disaggregation, may in fact be observed.

In fact there is an important disanalogy between economics and linguistics that will strengthen part of Nelson's conclusion. Transformational linguistics and psycholinguistics are both sciences of the individual. Once the former theory and the latter are linked in the explanation of individual speech output, the story ends. In the case of economics, there is a further stage, in fact the one that interests economists the most: how the individual choices add up and change as prices change, for example. The economist is committed to preserving the downward-sloped demand curves as reflecting the competence of an exchange economy as a whole, independent of its actual performance under governmental and other interferences. As we shall see in the next chapter, economists are equally committed to seeking equilibrium proofs as explanations of aggregate economic phenomena. If the performance theory to be harnessed with rational choice, now treated as a competence, does not preserve these results, it will be rejected, no matter what its value for the explanation of individual economic choice. In this sense, viewed as competence theory, microeconomics may be a stronger constraint on what permissible performance theories of economic choice will be like than transformational grammar theory is for psycholinguistics. Such a constraint is yet another reason why economic theory should appear completely impervious to empirical testing, when it is only just very well insulated from it.

As Nelson recognizes, the prospects for treating rational choice theory as a competence theory turn on the provision of a performance theory to link it with actual behavior. But before investing research in the development of such a theory, one needs some reason to think that the competence theory reflects real causal constraints on performance. Thus, in the case of linguistics, we can be fairly confident that some sort of axiomatization of a transformational grammar is realized in the head just because of the evident behavioral capacity to encode and decode an indefinitely large number of sentences. About the only way a finite brain could perform any one of an infinite number of tasks is if it did so recursively. So, the data in psycholinguistics make a transformational grammar of some kind a fairly strong assumption. Here there is a fairly direct relation between a large range of behavior and the competence postulated and made a subject of theorizing.

The same may not be the case with economics and psychoeconomics, however. There does not seem at the outset as much about actual human behavior crying out for a utility function and a set of expectations that operate together the way rational choice theory demands. There is introspection and folk psychology, to be sure, and there is the normative role of rational choice theory. These make a competence theory of expectations operating over preferences attractive. However, these considerations are nothing like so powerful as the evident behavioral fact which motivates the search for a transformational grammatical competence, and they provide far fewer evidential constraints on a rational choice theory than linguistic behavior places on transformational grammar. Finally, the linguistics/psycholinguistics case reflects the competence/performance approach in one fairly restricted and highly structured area of human behavior. Rational choice theory will be a competence theory for the entire range of human action. Given this difference, the prospects of a manageable performance theory that will convert its capacities into explanations and predictions of actual behavior must be far slimmer than those of psycholinguistics. After all, the list of the principal factors that will interfere with our production of grammatical sentences must be far smaller and easier to identify than the list of such factors that interfere with our making the most rational choice given our preferences and expectations. But this list is what the performance theory is supposed to provide. My guess is that this list of factors is so long, so heterogeneous, and so unsystematizable, that the competence/performance distinction is doomed to purely academic interest in the case of economics.

There may be a more serious problem facing microeconomic theory even when construed as a theory of competence without payoff for predicting and explaining actual choices of actual agents.

ARE THERE EXPECTATIONS
AND PREFERENCES?

There is at least considerable reason to think that as the brain is organized it is incapable of literally registering the sort of discrete propositional thoughts that constitute either the beliefs and desires of folk psychology or the expectations and preferences of economics. If this is so, then the attribution to the brain of expectations and preferences and of operations over them eventuating in actual choices (when mediated by performance factors) will be problematic from the point of view of psychology. For economics, the problem may be more serious, for if human brains are unable to realize the competence hypothesized by rational choice theory so interpreted, then when we add together the actions of individuals, we simply multiply the errors introduced by the false competence theory. If at best, rational choice theory is a useful fiction, a convenient instrument, or the description of behavior that actual choice approaches asymptotically, as the best-case scenario would have it, combining the explanatory or predictive output of a large number of such useful fictions must perforce result in disconfirmed claims about economic aggregates.

It is hard to cast doubt on the psychological reality of beliefs and desires. Introspection reports them and commonsense explanation has appealed to them as far back as the beginning of recorded history. It is easier to cast doubt on their causal role in the etiology of human behavior. And in doing so, we need not deny that our behavior is a fine-grained response to environmental stimuli—both immediate and mediate—which reflects goals, purposes, ends, and so on. What I deny is that the intervening variables linking goals and stimuli to behavior are propositional organized mental states.

To see why we should deny this, consider first of all the improvement problem with which we grappled in a preceding section. There we noticed that a psychological state like expectation or preference (like their more familiar cognates: belief or desire) cannot be characterized by descriptions of bodily movements alone and that such states are defined only by appeal to other psychological states. These psychological states are known among philosophers as *intentional* states. 'Intentional' has a special meaning in philosophy, though one which is related to its ordinary meaning of "purposefulness." To say that a state of mind, like belief, for example, is intentional in the philosopher's sense is to say that it has "propositional content"—that beliefs "contain," in some sense, statements. Thus, there cannot be a belief without a statement believed. Belief, it is often said, is a relation between a sentient creature and a statement: x believes that p.

The problem with this approach is how the physical matter in our brains can "contain" propositions that reflect our expectations and our preferences and how it can encode and retrieve such statements and make them causally effective in bringing about behavior. The problem of how the brain contains thoughts is crucial in the present context, because if this is something the brain cannot even in principle do, then a theory of competence for rational choice is doomed from the start. It may continue to play a role as an instrument for commonsense predictions and explanations, but as a theory which we expect to link up with considerations about performance, it must be surrendered. If the brain is in principle incapable of realizing the competence in question, there is no way causally to link it up with other factors that will jointly generate performance.

It is clear that the sentences contained by intentional states cannot simply be written on or in the synapses of the brain. Why not? Because to take this claim literally seems to involve an absurdity. Consider a library card catalogue. The ink marks on each card in the catalogue represent a book. But they do so because there are library users who *interpret* the ink marks. These ink marks and the cards they are on do not intrinsically and directly represent anything; they are just pigment and pieces of wood pulp. It is perfectly conceivable for such ink marks to have been formed on pieces of wood pulp by accident, in the way that a mountain cliff might resemble a human profile or a tree branch might have the shape of the letter Y.

Only sentient creatures who interpret the ink marks as having a meaning can give the cards and the ink marks their character as representations of something else, books elsewhere in the library. Now consider how the gray matter can represent a state like the expectation that the price of a good's complement has risen or the preference of one state of affairs—say, sunny and mild weather—over another—say, a snowstorm. When statements like these are represented in one part of the brain, call it l_1 for "library," who is the *interpreter* who treats the configuration of synapses at l_1 as expressing this statement? Surely it must be some system in the brain, call it l_2 for "librarian," which "reads" the statement—expressed in some neural code (on analogy with the genetic code of the DNA perhaps) —off the synapses. If we postulate such a subsystem, we must face the same question all over again for l_2, the part of the brain that does the reading. For it to read, it must represent somewhere within it the meaning of what it reads in the synapses at l_1. Let's call this part of l_2, which represents the statement that l_2 reads off the library l_1, the subsystem l_3. But the same question arises for l_3. How does it contain the statement that it reads off the larger subsystem of which it is a component? The result, instead of explaining how the brain can "contain" statements, is a vicious regress.

Surely there are alternatives to this view. Yes, but they have equally unpalatable consequences. One of them, associated with the philosopher John Searle, is simply to assert that the brain can and does contain statements, and that there is simply no explanation for how it does so. This view, which accords the brain "intrinsic intentionality," has all the advantages of simply pretending that the problem does not exist. If intentional states of the brain have causal consequences for behavior, there must be a mechanism whereby they bring about these effects. To assert baldly that they do and leave the matter at that is simply the repudiation of science, especially in view of the prima facie difficulty described above. Moreover, simply maintaining a dignified silence about how the brain represents will not do for present purposes, because without such details, there is no way systematically to link the competence theory of rational choice with a performance theory of factors that interfere with it to produce actual behavior that is imperfectly rational. How can one causal factor interfere with another unless the second can be deflected from its target?

The leading approach to explaining how the brain can contain statements is in effect a tacit admission that it cannot, or that if it does, the statements it contains cannot be separated from one another in the way they need to be in a competence theory or any theory that will improve on common sense in the explanation of behavior. According to this approach, widely known as "naturalism," the neural states that intervene between environmental stimuli and eventual behavior have their content by virtue of the appropriateness of the behavior to the stimulus.

Naturalism is easier to explain in the case of animal behavior than human action: suppose a mouse comes within the visual field of a cat, which watches it briefly and then pounces on it. Surely there is some neural state which contains the information that a mouse is present. Why?—because reading back from the pouncing behavior and what we know about the preferences of cats, the best explanation for the behavior is the content attribution. So, having a certain statement as a neural content is just behaving in the way appropriate to the truth of that statement. There are serious problems with this approach.

Exactly what "statement" are we to attribute to the cat's brain under the circumstances: presence of member of species *Mus mus?* appearance of natural prey? source of delicious dietary protein? this is a brown furry creature with whiskers? Nothing in the cat's behavior, then or later, will enable us to decide among these and a dozen more candidates for the "content" of the cat's brain. What is worse, attributing any of them involves crediting the cat with a conceptual scheme that includes the concept of species or prey or protein or whisker, etc. And none of these concepts seem reasonable to attribute to a cat. One might suppose that this problem does not arise with people, since their behavior is rich enough to enable us to more finely discriminate what neural content best explains their behavior, and besides, people have language so they can tell us exactly which proposition they believe. These differences are illusory. To begin with, as we noticed previously, any human behavior can be the product of an indefinite number of combinations of expectations and preferences, so unless we can be certain what individuals want, we cannot be certain exactly what propositions their neural states contain, and to be sure what propositions they want to be true requires us already to have fixed the propositions they believe. Moreover, it is probably the case that people's beliefs and desires are not fixed with the precision that their propositional descriptions appear to provide. The specificity of descriptions of beliefs and desires is a function of the fact that we use language to express them; and this specificity is probably an artifact of our use of language to do so.

The upshot is that there is almost as much indeterminacy in attributing human beliefs and desires on the basis of behavior as there is in attributing cognitive states to animals. There is just too much slack here to treat attributions as more than just guesses that beyond certain limits, long ago attained in commonsense explanations, we cannot improve. As we have seen, there is no way to decide what exactly an individual really believes or desires, because an indefinitely large number of different packages of belief and desire pairs will generate the same behavior. If this is so, then in principle two lifetimes of exactly the same sequence of actions could be the result of two different collections of expectations and preferences. The only way to distinguish the causes of the actions in these two lives would be to examine the heads of these two agents.

But this is where we came in, so to speak. The aim of widening the focus from the brain state itself to the stimuli in the environment and the behavior that eventuates is so that we can read content in. If we need the content in order to distinguish one package of beliefs-cum-desires from another, our problem has not been solved.

In the face of this problem one solution we cannot try is that of giving up the idea that preferences and expectations contain statements. Indeed, this is how we distinguish one belief from another. Thus, the difference between your belief that $4^2 = 16$ and your belief that $5^2 = 25$ is given by these two different sentences that your beliefs "contain." And presumably your belief that the price of tea will remain the same tomorrow is the same as my belief that it will, because the two statements they contain are identical. Preferences are also identified by the statements they contain: my desire for a car is the desire that I have one; my preference for A over B is my desire that I receive A when both are available.

Philosophers have a technical term for the fact that desires' and beliefs' identities consist in the statements they contain. They refer to such states as having "intensionality." This is a notion easier to illustrate than to explain. Take the sentence "Lois Lane believes that Superman is brave." Now, Superman is identical to Clark Kent. So, if we substitute equals for equals in the description of Lane's belief, we turn a truth into a falsehood, viz., "Lane believes that Clark Kent is brave," for Lane believes quite the contrary of Kent. But compare "Superman was born on Krypton." If we make the same substitution, we preserve the truth in the resulting sentence: "Clark Kent was born on Krypton." Thus belief and desire sentences are sensitive to substitutions of coreferring terms and to coextensive predicates for that matter.

Another way of putting it is to notice that Lois Lane has a belief about Clark Kent's being brave, but not "under that description." This sensitivity of intentional states to the descriptions and terms we use to identify them is even clearer in the case of desires. Thus, "Lois Lane wants to marry Superman" is a true statement about that newspaper woman's desires. Although Superman is identical with Clark Kent, it is plain that Lois Lane does not desire Kent in marriage. Lois Lane wants to marry Clark Kent under one description, but not under another.

So, intentional states are ones in which we cannot freely substitute synonymous descriptions without risking changing a truth to a falsity. Philosophers have a special name for this feature of intentional statements. They call them "intensional"—with an s instead of a t. The intensionality of our descriptions of beliefs and desires shows two things: first, no alternative means will ever be found to identify and measure them, and second, their identities, their characters, are given "holistically." Thus, they are forever insulated from the sort of factors that a nonintentional performance theory might cite to convert rational choice capacities into actual behavior.

Consider the first point. There are only two sources for a determination of what someone believes or desires: behavior or brain. What we need is something that will "measure" what a person believes by some distinct effect of the belief, in the way that a thermometer measures heat by its quite distinct effect, the height of a column of mercury or alcohol. We need an equation with a belief (or a desire) on one side of the equal sign and a brain state or description of behavior on the other. But such an equation is impossible; something will always be missing from the brain or behavior side of the equation: intensionality. The description of behavior or brain states is never intensional. It is always extensional: any true description of a bit of mere behavior or of a brain state, whether in the language of anatomical displacement, physiology, cytology, molecular biology, chemistry, or electromagnetic theory, will remain true whenever we substitute equivalent descriptions into it.

Consider the second point. Sometimes we can substitute equals for equals in belief and desire descriptions—coreferring terms and coextensive predicates—while preserving the truth values of these descriptions and even preserving the belief or desire referred to by the whole sentence. Thus, your belief and mine that Superman is brave are most probably the same belief as the belief that Clark Kent is the opposite of cowardly. But the reason is that we have certain other beliefs, for example, the beliefs that Clark Kent is identical to Superman and that cowardice is the opposite of bravery. What this shows is that the identity of any one of our beliefs and desires is a function of what other beliefs and desires we have. Accordingly, to establish very firmly that an agent has a particular specific belief and/or desire, we need to know a vast amount about what other beliefs and desires the agent has—whence the "holism of the mental." That is, mental states, it seems, cannot simply be divided up and counted out. What we believe and desire is a whole not uniquely divisible into discrete parts. Thus the mind is inaccessible to a simple theory of rational choice competence.

Notice that under the conventional assumption of perfect information, both about alternatives expected and about one's preferences, the propositional character of preference and expectation is obscured. Since the agent knows all the relevant identities and coextensivities, he or she will never make a mistake. However, these conditions never obtain, and even if they did, the agent's choices would still be determined by contentful states we could not infer from choices or brain states.

Of course, one can claim that every particular instance or token of an intentional state is identical to some particular instantaneous brain state or other. But this is no solution to our problem, either theoretically or practically. To begin with, the claim that brain states encode statements we believe threatens psychology with an infinite regress and can provide no explanatory basis for rational choice theory. Consider, if one part of the brain is to store statements, then another part must read and interpret the statements the brain "encodes." But this introduces the problem of intentionality all over again. For one part of the brain to interpret data in another part, it must express and employ statements describing the data. How does the interpreting part of the brain do this encoding? If the answer is that it does it the same way the storage system does it, we are off on an infinite regress without explanatory impact. This token identity does not enable us to identify the type of belief or desire that any particular brain state realizes.

More practically, to secure an independent nonintentional measure of repeatable beliefs and desires that can figure in the explanation and prediction of behavior, we need to link types of mental states with types of brain states. This is just what intensionality and the slack in reading content back from behavior show us is not possible. In fact, it will turn out that since what I believe is a function of all or most of my other beliefs and desires, then any one of my belief states is identical to the state of a big chunk of my whole cerebrum. But no description of that much of my brain would provide a useful way of measuring my expectations and desires.

One problem, which need not concern the philosopher of economics, is where this diagnosis leaves folk psychology and commonsense explanations of human action. Briefly, to the extent that concepts like belief, desire, and their cognates are employed to systematize behavior, they function as part of a heuristic device, an instrument useful up to a point in organizing our everyday experience, but incapable of improvement beyond the predictive powers reflected in our every projection of people's actions. At best the brain asymptotically approaches the kind of contentful representations folk psychology requires, enough at any rate to make introspection seem like a credible source of insight into the causes of our own behavior. But neither introspection nor observation of the sort everyday life provides can underwrite an improvable science. Of course, science is not what we need for everyday life.

Can economics help itself to the same rationale? Only if its aims are akin to those of getting along in everyday life. If we took the self-denying ordinances of McCloskey seriously, we should in fact set the aims of economics at the same level as those of folk psychology. McCloskey tells us that economic theory's best answer to the "American question: if you're so smart why aren't you rich" is that it isn't any smarter than well-informed common sense.[11] But it is evident that neoclassical economics has higher goals than these.

11. *If You're So Smart* (Chicago: University of Chicago Press, 1991).

CONCLUSION

Now, at any rate, we have an explanation for why the assumptions of economic theory about individual action have not been improved, corrected, sharpened, specified, or conditioned in ways that would improve the predictive power of the theory. None of these things have been done by economists because they cannot be done. The intentional nature of the fundamental explanatory variables of economic theory prohibits such improvement.

The upshot is that the sort of research program Nelson envisions to link up microeconomics with something that could make it an empirical science is not in the cards—not because the venture is logically impossible but because the variables of folk psychology, which economics has embraced and formalized, just cannot be linked up with the rest of science and cannot even be realized, exemplified, or instantiated by the brain in theoretically tractable ways. The prospect that we cannot link intentionality up to the sciences is one that Nelson recognizes and one whose consequences he briefly considers:

> If the program I have defended turns out to be an empirical failure, it would still have important and instructive consequences. . . . [W]ith respect to economics, such a failure would provide strong evidence, perhaps overwhelming evidence, against the possibility of individualistic foundations for microeconomics and, hence, for economics in general. Since economics is such a highly developed social science, this would in turn create a strong presumption against the possibility of individualistic foundations for any social science. (P. 489)

How "instructive" for economics these consequences have been, or ever will be, is a moot point. Although it is hardly news that the discipline's theory of human choice has been an empirical failure, its treatment as a failure cannot after all plausibly be traced to the mistake of confusing a competence theory for a performance theory. And we can be sure that economists will remain as indifferent to the findings of psychologists (behav-

ioral, cognitive, or otherwise) as they always have been. This indifference is a matter to be dealt with in subsequent chapters.

One question that seems hardly worth asking is, Why in the
face of the difficulties about its theory of rational choice and the
further problems it has not even recognized does the discipline
not simply abandon rational choice and give up the individu-
alistic foundations to which Nelson refers? One part of the
answer is the discipline's commitment to "methodological individualism," the view that large-scale economic processes are to
be explained by appeal to aggregation of the behavior of individual agents. This thesis, however, is not in question here,
since methodological individualism does not require that individual behavior proceed in accordance with rational choice
principles. So far as it is concerned, choices can be the result of
any process whatever. Thus, economics can surrender rational
choice theory while continuing to honor the strictures of methodological individualism. In this respect, Nelson's conclusions
are too pessimistic. On the other hand, it is difficult to envision
a nonintentional theory of individual behavior. Such a theory
would be unable to take behavior seriously as action, and it could
not be linked up with the whole scheme of legal and moral responsibility that folk psychology underwrites. Even theories that
proclaim themselves free and independent of the snares of folk
psychology find themselves implicitly reintroducing the very
same kind of intentionality with which it is burdened.[12] Even
theories like Freud's that find the causes of much behavior in
unconscious forces still treat these forces as desires and beliefs,
albeit normally inaccessible to introspection.

To deny that individuals have desires and beliefs seems incoherent. After all, denial is impossible where there are no beliefs and desires: to deny something requires that you believe it
is false and desire to say so. But for a theory to ignore preferences and expectations in the causal explanation of behavior is
tantamount to denying them existence. It relegates preference
and expectation to the rubbish bin of science, along with epi-

12. This is particularly true for behavioral psychology. See Alexander
Rosenberg, *The Philosophy of Social Science* (Boulder: Westview Press, 1988),
chap. 3.

cyles, philosopher's stones, and phlogiston, things whose existence we deny because they have no role in our best scientific theories. The conclusion that beliefs and desires don't explain behavior just seems inconceivable. Introspection alone assures us that our actions are caused by our beliefs and our desires, and no amount of predictive failure on the part of a social scientific theory is going to shake this conviction, among ordinary people and among economists. Accordingly beliefs and desires, preferences and expectations, will always be with us. On the one hand, proposals that we surrender intentional microfoundations for economic theory had better reconcile themselves to this fact. On the other, economists committed to treating their discipline like a scientific theory had better recognize the limitations under which their preferred variables place them.

6

COULD ECONOMICS BE A BIOLOGICAL SCIENCE?

Economists have long been cognizant of the intractable character of their explanatory variables, both preferences and expectations. Although they have not explicitly recognized the source of their problems in the intentionality of preferences and expectations, they have taken steps that effectively neutralized the impact of these problems of intentionality on the exposition of economic theory. But the cost has been high. Neoclassical theory long treated expectations as implicitly extensional by assuming that economic agents had complete information about available alternatives. This omniscience effectively means that agents have only true beliefs, so that substitution of coreferring terms or coextensive predicates in descriptions of their beliefs always preserve their truth. Assuming agent-omniscience seriously weakens the power of economic theory. Allowing for intensionality is in effect to recognize that agents are not omniscient—sometimes they have false beliefs or no beliefs about certain states of affairs. However, as I argued in chapter 5, accommodating intensionality seems impossible in a predictively improving scientific theory.

Similarly, preferences have been recognized as recalcitrant to economic theorizing, though again not because of perceived problems of intentionality. Over the course of a century, from marginalism through revealed preference theory, economics has witnessed a series of retreats on the part of economists from quite substantive claims about tastes to a discrete silence.[1]

---

1. I catalogue some of these shifts and offer an explanation for them in "A Skeptical History of Microeconomics," *Theory and Decision* 12 (1980): 79–93.

Tastes are intractable in ways well known to economists. When described in terms of utilities, they are neither intrapersonally comparable with respect to cardinality nor interpersonally comparable ordinally. Nor is there in economics or elsewhere anything remotely like a theory of how they are determined; more important, there is no theory of how they differ between agents and how they change over time. Neoclassical microeconomic theory circumvents these problems by treating tastes as "exogenous." They are determined by noneconomic forces and are "given" for the purposes of explaining and predicting behavior. They are subject only to ordinal intrapersonal comparisons and sometimes are treated as mere notational variants on actual choice behavior, systematized in the theory of revealed preference. When so treated by economics, economists deprive themselves of the resources for explaining individual choice and actual behavior, since preferences cannot explain what they are when treated as mere notational variants of actual behavior. Agents' preference structures and changes in them are never the effects of changes in prices and quantities of commodities available; rather tastes are (among) the noneconomic causes of price and quantity produced. Treating them exclusively as causes is in many ways a reasonable approach to take toward preferences, for there are good grounds to suppose that tastes and preferences are formed by noneconomic forces. On the other hand, the approach seriously compromises the explanatory and predictive power of economic theory and does so for reasons beyond those canvassed in previous chapters, as we shall see.

For welfare economics, the consequences of the intractability of tastes seem at first blush more serious and certainly more obvious. The limitations on normative conclusions about welfare that are enshrined in the criterion of Pareto optimality testify to the effects of enforced agnosticism about tastes. Because tastes are exogenous, there is no scope within welfare economics for even so elementary a distinction as that between wants and needs (beyond an altogether inadequate account in terms of differences in elasticity). Moreover, considerations of Pareto optimality cannot distinguish between societies characterized by widespread slavery or starvation, as opposed to affluent and democratic ones, just so long as, in each of these three sorts of societies, no one can be advantaged without disadvantaging someone else. The reason is of course that within these societies the preferences of agents cannot be compared, weighed, adjudicated, or traded off against those of others. Tastes are incomparable. There is no disputing them. More generally, if the socially preferred alternative is not to be entirely independent of individual preferences, its selection will be no more determinate than the individual preferences it reflects.

In the first five sections of this chapter I will explore alternative approaches to preferences and expectations, desires and beliefs, that might afford alternatives to the conclusions of chapter 5, alternatives that offer more optimistic prospects for the predictive improvement of economics or that at least circumvent the theoretical problems generated by the intentionality of economic theory's explanatory variables. Mainly I focus on an approach to preferences which claims to avoid the methodological difficulties identified above and which additionally would eliminate the role of intentionally characterized desires altogether. Exploring this theory will help us see how serious the problems of intentionality are and the lengths to which one must go in coping with these problems. Moreover, the theory I examine is a promising one, preserving much of the systematic achievement of the theory of consumer behavior while holding out the hope of fertile and fruitful development. It also has grave limits, which need to be clearly identified. The theory in question is due to Gary Becker.[2] Becker's "new theory of consumer choice" aims explicitly at solutions to these two problems of the exogeneity of tastes and the weakness of welfare conclusions. Moreover, it also holds out the hope of providing a nonintentional approach to preferences, as we shall see. As such, it is halfway toward circumventing the obstacles to predictive improvement of microeconomic theory identified in chapter 5.

I shall argue that Becker's innovation holds out hope of circumventing *some* of the problems of intentionality canvassed in

2. *The Economic Approach to Human Behavior* (Chicago: University of Chicago Press, 1976). All page references to Becker are to this work. See also Becker's *A Treatise on the Family*, 2d ed. (Cambridge: Harvard University Press, 1991).

the last chapter only if it is interpreted as a theory about the *biological* needs of *Homo sapiens*, instead of a theory about the preferences of economic agents. This is an interpretation which Becker himself encourages, though he does not explicitly adopt it. Although substituting biological needs for preferences will not in the end transcend the limitation to generic prediction of economics, the hope that "biologizing" economic theory will do so is a long-standing temptation in the neoclassical tradition within which Becker was trained. In "The Methodology of Positive Economics" Friedman offers a rationalization for neoclassical theory that trades on parallels with evolutionary biology. In the last several sections of this chapter I examine economists' arguments for this parallel, largely to show that it provides little grounds for optimism about economic theory's power to move beyond generic prediction.

THE PROBLEM OF EXOGENOUS TASTES

Becker's theory is the most important, influential, and empirically powerful reformulation of the theory of economic behavior offered since ordinal utility replaced marginalism. Moreover, Becker's work has spawned a movement in economics called "human capital theory" that comes close to being a revolution in economics. It has had great influence on the economic treatment of public policy. In addition it has vastly broadened the applicability of economic theory to a large number of apparently noneconomic choices. Among those areas to which the new theory has been applied with impressive empirical support are marriage, family size, divorce, travel, education, migration, health, and "cross-sectional and life-style patterns of consumption expenditure and time allocation." Indeed, Becker holds that his "approach provides a valuable unified framework for understanding *all* human behavior" (p. 14). I think this claim is correct. The theory has far greater explanatory power than traditional microeconomics. But in this chapter I shall ignore its powers of deductive systematization of independently generated data and focus on its foundations.

Economics, as we have seen, is an intentional science. It holds that economic behavior is determined by tastes and beliefs, that is, by the desire to maximize preferences, subject to the constraint of expectations about available alternatives. Differences between the choices made by individual agents when faced by the same alternatives are due to either differences in preferences or differences in expectations or both. Similarly, changes in the choices of an individual agent over time are due to changes in one or both of these causal determinants of behavior.

I have argued that these intentional variables are intrinsically inaccessible to the sort of regimentation that would produce an improvable scientific theory. And as noted above, to an extent, the reasons for this inaccessibility are already recognized by economists. According to Becker, the methodological problem for economics is that there are no resources within its theory for separating the influence of these two components on changes and differences in actual choice. (Hereafter I will speak of differences, meaning thereby both intrapersonal changes in preference and interpersonal differences in it.) Yet, he notes, such separation is required to preserve microeconomic theory against the joint charges of triviality or falsity. Actual economic agents often enough seem to violate the principles of rational choice, for example, by making purchases at different times that violate the weak axiom of revealed preference. Does this show that the theory is false? No. One apparently reasonable thing to say when an economic agent's behavior appears to be irrational because it violates the preference ranking previously assigned to him is that his tastes have changed. But if tastes are exogenous and nonfungible, there is no prospect of an economic or indeed any other systematic account of such changes, and so no prospect of assessing this strategy for explaining away apparent irrationality. The price of theory preservation here seems to be triviality.

Becker expresses this problem in the following terms:

For economists to rest a large part of their theory of choice on differences in tastes is disturbing since they admittedly have no useful theory of the formation of tastes, nor can they rely on a well-developed theory of tastes from any other social science, since none exists. . . . The weakness in the received theory of choice, then, is the extent to which it relies on differences in tastes to "explain" behavior when it can neither explain how tastes are formed nor predict their effects. (P. 133)

The problem runs very deep in economic theory. Most of the interesting theorems of microeconomics involve holding either or both preferences and expectations constant, while varying prices and/or income. Expectations are held constant by the assumption of perfect information. When this assumption is relaxed, expectations turn out to be a function of the agent's tastes and probability assignments. Preferences are held constant by assuming that the individual remains on the same indifference curve or in the same family of them.

As Becker notes, there is at present no theory elsewhere in social or behavioral science that treats tastes as endogenous to which economics can help itself or that can at least comfort economists with the assurance that an account of changes in taste is, at least theoretically, available. Economists may, however, respond to a charge like Becker's with the claim that such a theory is, in principle, possible. Meanwhile, economists can certainly appeal to tastes as a legitimate explanatory variable in their theory. Solving some problems about human behavior need not require that we solve all problems, especially when these other problems are the business of psychologists and not economists.

The trouble with this line is that psychology seems incapable in principle of providing a theory of tastes that does not already presuppose the sort of economic rationality which economists need a theory of tastes to defend. As I tried to indicate in the previous chapter, much recent work in the philosophy of psychology suggests that beliefs and desires cannot be separated. This suggestion, often called "the holism of the mental," asserts that the identity of a preference or a belief is a function of the other beliefs and desires the agent has. (It is ironical that the impetus of this thesis is to be found in the same work of Frank Ramsey to which so much of the economic theory of choice under uncertainty is beholden.)[3] If beliefs and desires are so tightly connected, then we can never hope to separate tastes from expectations in such a way as to legitimate the appeal to them to explain away choices that falsify the theory of consumer behavior.

We may illustrate the problem that holism makes for economic theory by considering the well-known Slutsky equation, which expresses the separate effects of a change in prices and income on the expenditure of a consumer. For example, a sharp price drop in one commodity can affect the consumer in two ways: she can now buy more of the good, and she need spend less to purchase the same amount of it. Indeed, she may find herself so much better off by the price drop that she can afford to buy a good she preferred all along but could not afford. If this good substitutes for the cheaper good, the consumer may buy less or none of a good whose price has fallen, instead of more. Which of these outcomes will be the case is systematized in the Slutsky equation:

$$\frac{\delta q_i}{\delta p_i} = \left(\frac{\delta p_i}{\delta p_i} \right)_{\text{utility = constant}} - q_i \left(\frac{\delta q_i}{\delta y} \right)_{\text{price = constant}}.$$

The first term describes the rate of change in consumption of commodity i, q_i, for a change in its price, p_i. The equation attributes this effect to two causes: the first is described in the equation's second term as the substitution effect, the rate of change in consumption for a change in price (holding utility levels, i.e., tastes, constant). The second is the so-called income effect, described in the third term, which reflects how the change in i's price affects the agent's income, y, and through it her consumption of good i. If the second term exceeds the third, the consumer buys more of i as the price drops; if the third term is larger than the second, she buys less. But how are we to ensure that utility level is being held constant through such price changes? Ensuring the constancy of utility levels appears to be a question for the psychometrics of taste and prefer-

---

3. See R. B. Braithwaite, ed., *The Foundations of Mathematics and Other Essays* (London: Routledge and Kegan Paul, 1930).

ence. The only way the psychologist can measure preferences is by assuming that the subject is rational and varying the subject's beliefs about the costs and payoffs of different alternatives presented to him. In other words, the psychologist must adopt the assumption of utility maximization from the economist.

In fact, the psychologist may be faced with experimental data that can themselves be explained only by the Slutsky equation, thus bringing the inquiry back to where it started. Suppose the psychologist varies prices and the subject's income in order to see whether choices vary systematically in a way that reflects constancy of preference. If some of the goods whose prices are varied are inferior, then preferences will *appear* to be symmetrical as prices change, and the agent's choices will sometimes seem irrational: the consumer will prefer good i to j, and j to i. Either that or her tastes really have changed. Of course, the Slutsky equation can be invoked in these cases to explain away the apparent irrationality, as due to the income effect of an inferior good, without appeal to exogenous changes in taste. But this equation is the very one we are attempting to assess while holding utilities constant. If the philosophical analysis of the identification of psychological states is right, there is no alternative to this sort of circularity. Any actual choice can be accommodated to the conventional theory of consumer behavior because of the indeterminacy of the preferences that give rise to it. Even holding expectations constant through the assumption of perfect information will not take up enough slack, for there is no way to pin down tastes within the theory. Moreover, there seems to be no other theory in which economically exogenous tastes are endogenous. There is no prospect of a psychological theory of preferences that will enable us to identify changes in them independently of the identification of beliefs. The lack of such a theory is the source of the apparent empirical emptiness of the conventional theory of consumer behavior. Therefore, matters are worse than Becker supposes. It is not just that there is no psychological theory of tastes independent of the assumptions of economic theory; there cannot be one.

One traditional reaction to these problems about the theory of individual economic behavior is to deny that economics really has anything to say about individual behavior. We have noted such denials before and will come up against them again: claims about individual agents are treated merely as convenient fictions helpful in systematizing the real subject of economic concern: markets, industries, and whole economies. Some economists have followed Sir John Hicks's strategy for dealing with the Scylla and Charybdis of falsity or tautology that seems to face the theory. Hicks wrote that "if our study of the individual consumer is only a step towards the study of a group of consumers, these falsifications can be trusted to disappear when the individual demands are aggregated."[4]

Becker has adopted a similar view himself:

> Although economists have typically been interested in the reactions of large markets to changes in different variables, economic theory has been developed for the individual firm and household with market responses obtained simply by blowing up, so to speak, the response of a typical unit. Confusion results because comment and analysis were directed away from the market and toward the individual, or away from the economist's main interest. (P. 154)

The attitude toward the conventional theory of consumer behavior reflected in this passage is a coherent one. But it incurs considerable costs, costs so great that few economists are willing to pay them, including those like Becker who seem to endorse this strategy. One cost is that the theory of consumer behavior, treated now as a *façon de parler* for downward-sloping demand curves and without any real significance for actual individual behavior, loses its considerable powers to systematize that behavior. Thus, for example, despite its problems, the Slutsky equation really does seem to be an important part of the correct explanation of why the consumption of bread declines when its relative price falls greatly and why this phenomenon is not incompatible with the more frequently observed phenomenon of increased consumption in the wake of a price decline. A more serious loss is that this interpretation of microeconomic theory deprives it of its relevance to the whole domain of welfare economics. In this domain, the chief questions surround the effects

4. *Value and Capital* (Oxford: Oxford University Press, 1939), p. 23.

of social policies on the utilities of actual individuals, rather than on abstractions convenient for the economical expression of regularities about markets and industries. There would be no point to various compensation criteria for changing distributions of commodities, nor any sense in debates about the losses or gains involved in the existence and distribution of a consumer's surplus, unless there were real agents to be advantaged or disadvantaged by these distributions. The very possibility of normative applications of microeconomic theory depends on its actual applicability to real economic agents. And this application is seriously compromised by the lack of a theory of tastes.

BECKER'S NEW THEORY
OF CONSUMER CHOICE

Becker's new theory circumvents this problem of tastes entirely by the simple expedient of assuming that every agent has exactly the same set of lifelong tastes for the same set of quite specific goods as every other agent. Becker's theory involves an imaginative rearrangement of concepts familiar from conventional microeconomics. In particular it involves adapting the theory of production (where questions of taste do not arise) to the choices of individual agents. Instead of assuming that the agent's utility is directly derived from the consumption of market goods, Becker assumes that it is only indirectly dependent on them. Utility is derived directly from commodities the agent produces for his own consumption by combining market goods together with his own time in accordance with techniques of production available to him. The consumer's demand for market goods is identical to that of a firm for inputs or factors of production. Relatively simple mathematics illustrates these relations. Utility is a function of household commodities only:

$$U = u(Z_1, Z_2, \ldots, Z_n), \tag{1}$$

where U is utility, and Z_i is the quantity of household commodity i. Commodity i is produced for direct consumption from goods bought on the market combined with quantities of the agent's time. The amount of i produced is a function of the market goods purchased, the time available, and the technique of production:

$$Z_i = z_i(x_i, t_i, E). \tag{2}$$

Becker calls E the environmental variable. It describes the productive process or technology which the agent employs to produce Z_i from x_i and t_i. As in the conventional theory the utility function (1) is maximized subject to the constraint of (2) plus a constraint on income and one on the household's available time. Total income, I, is all spent on market goods, without any savings. It is the sum of the amounts of market goods purchased times their prices:

$$I = \sum_{i=1}^{n} p_i x_i, \tag{3}$$

where p_i and x_i are the prices and quantities purchased of market good i. The total time, T, an agent disposes is the sum of the time he or she spends in the labor market to earn income, t_w, and the time he or she devotes to household production of Z_i, t_i:

$$T = t_w + \sum_{i=1}^{n} t_i. \tag{4}$$

Maximizing the utility function (1), subject to the constraints of (2), (3), and (4), enables us to derive conditions on optimization of consumer choice familiar from the theory of the firm. Thus, when utility is maximized, the ratio of the marginal products of any two factors of production (i.e., market goods) must be equal to the ratio of their market prices. Similarly, the ratio of the marginal utilities of two household commodities must be equal to the ratio of their marginal costs—their shadow prices for the consumer—determined by the cost of the market goods and time required to produce them and by their respective techniques of production.

Little has been said about E, the environmental variable of equation (2). The techniques of production it reflects will be crucial to the shadow prices of household goods and the productivity of market goods and time in producing them. An inefficient method of production will prevent the highest attainable level of utility for a given set of market goods and available time from being reached. In fact E will have to take up the work that tastes are no longer able to do in this theory. Interpersonal and intrapersonal differences in the consumption of market goods will have to be accounted for by differences and changes in E.

Now, the chief virtue that Becker claims for this reformulation of consumer behavior theory is "its reduced emphasis on the role of 'tastes' in interpreting behavior." He recognizes that the "shift in emphasis towards changes in prices and income and away from changes in taste may appear to be simply one of semantics—of hiding an inability to explain tastes behind the camouflage of a production function" (p. 144). Becker's explicit response to this charge that our knowledge of the quantifiable properties of production functions employed by agents can help us reduce the need for appeals to changes in taste is inadequate.

First of all, the problem is not that the old theory's scope for changes in exogenous tastes is too large. It is that we can never tell how large or small it is. We still cannot tell how large it is, even with Becker's new theory. Second, substituting household commodities for market goods in the utility function by itself does nothing to reduce the scope of tastes at all. In principle, consumer tastes for household commodities can change as easily as tastes for market goods. Third, defenders of the conventional theory can reduce the role of tastes as effectively as Becker can, simply by weakening the assumption of perfect information and raising the costs of acquiring information. Indeed, various formulations of the conventional theory have done this very thing. The tactic of weakening the assumption of perfect information of course does not increase the empirical content of the theory, because there is no available measure of an agent's information or how he or she exploits it, for the same reason that tastes are intractable. If the new theory is to have a real methodological advantage over the old theory, it cannot be this one. Rather, Becker is committed to a thesis that he seems only to entertain and not explicitly to endorse:

Consider a logical extension of the view that behavioral differences previously attributed to differences in taste are in fact due to differences in productive efficiency. One might argue that indeed all households have precisely the *same* utility function and that all observed behavioral differences result from differences in relative prices and access to real resources. In the standard theory all consumers behave similarly in the sense that they all maximize the same thing—utility or satisfaction.[5] It is only a further extension then to argue that they all derive that utility from the same "basic pleasures" or preference function, and differ only in their ability to produce these "pleasures." From this point of view, the Latin expression *de gustibus non est disputandum* suggests not so much that it is impossible to resolve disputes arising from differences in tastes but rather that in fact no such disputes arise. (P. 145)

Although Becker does not want to endorse this thesis fully, it seems evident that he is committed to it, and thereby committed to further conclusions about the structure of the preference function expressed in equation (1). Without it, there is no real solution to the problem of tastes for microeconomic theory, for we can no more easily separate intrapersonal and interpersonal differences in E, the means of production an agent chooses to employ from such changes, and differences in the set of household commodities he chooses to produce. Indeed, without this strong assumption of identity of utility functions, Becker's formalism is even more prey than the conventional theory to the exogenous taste problem, for to the degree an agent chooses productive techniques, his choice will be determined by his tastes for these techniques as well as his factual beliefs about them. Unless all tastes are the same so that they cancel out in an explanation of why different productive techniques are exploited, we have simply increased the problem of tastes by employing Becker's theory, not decreased it.

5. Becker probably realizes that this is a tendentious characterization of the theory of consumer choice. In light of the problems of revealed preference theory, this view of consumer choice seems warranted in spite of its character.

Indeed, Becker requires more than merely a "logical exten-sion" of the view that some behavioral differences attributed to taste differences are really due to differences in productive technique. It is a difference of kind from the old theory that he requires. To say that everyone always has the same utility func-tion is to say that each agent's utility will be maximized by the same bundle of the same amounts of a limited number of fun-gible but independent commodities (given the same income, time, and production function constraints, of course). Nothing less than such a utility function will do the job that the old the-ory fails to do. But, I shall argue, such a utility function will be a schedule of needs, not preferences.

To see that the number of household products must be fairly large and measurable, imagine that there is only one household commodity, say, good health (both physical and psychological), that all agents produce in order to maximize their utility. It is worth noting that Plato might well be said to have held this view. The trouble with this supposition is that a given amount of the commodity in question supervenes from many different com-binations of market goods. It can be produced by many alterna-tive techniques about which relatively little is known and is itself difficult to measure. Therefore, it bears all the problems of in-comparability and nonfungibility that utility does. A single, global, directly consumed household good that in effect does nothing but intervene between utility and market goods will be no improvement on the old theory, especially when the "pro-ductive techniques" for generating this good from market com-modities are so "subjective." To say that smokers and non-smokers have exactly the same tastes and maximize utility by maximizing their health either involves treating health as a mere notational variant on utility or raises questions about the efficiency of productive techniques that are as unanswerable as questions about the comparisons of utilities.

Suppose we increase the number of basic commodities to three, as Hobbes seems to have advocated in *Leviathan*.[6] We sup-pose that utility is maximized by producing as much "safety," "gain," and "reputation" as possible. Here again we have com-

6. Pt. 1, chap. 13.

modities which are scarcely easier to measure than utility. But waive this problem. If there are trade-offs between these or any household commodities that keep the agent at the same level of utility, the indeterminacy we had hoped to escape reappears. In such a case, interpersonal and intrapersonal differences in the production of household commodities may appear and may be optimal for their respective agents. But distinguishing such cases from irrational choices or ones dictated by different techniques of production is the problem of exogenous tastes all over again. We cannot tell whether an agent is indifferent between two different combinations of household commodities attainable by him or whether his tastes for these commodities have changed. If household commodities are subject to trade-offs, we have merely complicated the methodological problem of tastes, not simplified it. To implement the new theory beyond the limits of the conventional one, we shall have to identify a fairly large, varied, but manageable number of universally desired household commodities. Their amounts must be measurable and independent of one another in their effects on utility. Their productive techniques must also be discoverable, as well as quantitatively comparable for efficiency. What are the prospects for discovering such commodities?

STABLE PREFERENCES AS HUMAN NEEDS

The first thing to notice is the similarity of household commodities that meet these conditions with what we may independently identify as basic human needs. If we can parlay this similarity into something approaching an identity, then one of the most well known objections to economists' view of human behavior will have been circumvented. Microeconomic theory resolutely refuses to distinguish between mere wants or desires and needs. It does so of course because of the incomparability of tastes—of wants and needs. If some or all household commodities are those that meet human needs, then economic theory can take needs seriously, with potentially great advantage for welfare economics.

Why describe the demand for household commodities as needs? First, because it is obvious that individuals have a variety of different needs and that these needs cannot be traded off, and in many cases we know at least in principle how and in what units to measure them. We know also that few market goods are directly needed and most are purchased in order to meet needs—such as nutrition, warmth, shelter, amusement, and psychological well-being. Market goods are combined in a variety of different ways—by different techniques of production—in order to meet these needs. Another reason for substituting needs for Becker's uniform preferences is that this assumption makes the most sense of the obvious fact that although each person has different tastes, each person explains and justifies his or her tastes by showing how they answer needs, given constraints. Tastes turn out to be beliefs about how to meet needs. For example, the cultural anthropologist interpreting the desires and values of an alien culture will attempt to show how its values—so different from our own—emerge from that culture's attempts to meet needs common to all cultures. In the past anthropologists often accounted for these differences by citing fallacious beliefs held by natives, for example, by showing how inefficient their techniques of production were for meeting their needs. Nowadays, anthropologists steer clear of such ethnocentrism. Instead they attempt to show how ecologically adaptive native productive techniques are and how they often meet needs the agents themselves do not even recognize—optimizing family size, for example.

Finally, if we do not treat the utility function as one reflecting needs, the advantage of Becker's new theory is again jeopardized. To begin with, if we interpret household commodities in a literal sense according to which a household commodity is what each family cooks for dinner, chooses for its evening's entertainment, chooses for its frequency of bathing, and so on, the assumption of uniform and unchanging preferences is simply false. We shall then have to have recourse to all those dodges which the old theory needs to render its unrealistic claims consistent with the facts. If we treat household commodities as merely fulfilling preferences, interpreted now as psychological wants, we shall be no more able to identify the common household preference function shared by all individuals than the old theory is able to independently identify the idiosyncratic preference functions of individual agents. The inability to specify the preference function is admittedly less of a problem for Becker's theory, for by assuming that all individuals have the same preference function all the time, the new theory will never need to specify these preferences independently, since they never enter into an explanation of behavioral differences or changes. But this tactic in effect simply shifts the entire problem of preferences onto psychology, where it is no more tractable than when it seemed to be part of economics. The shift Becker proposes would not be "simply one of semantics," but it would be one of merely divisional redistricting or gerrymandering in the behavioral sciences. It would not solve a problem so much as ship it to another department.

For all its attractiveness, the treatment of the new theory's preferences as needs is not one Becker himself offers, of course, but it is one Becker's own speculations encourage. We already know a fair amount about the needs common and permanent among economic agents. They are specified in life science's findings about minimal levels of sustenance, the limits on life-sustaining climate and weather, and the other variables on which the persistence and evolutionary fitness of *Homo sapiens* depend. We know a great deal about these factors, and we can expect to come to know much more about them. Needs so understood are as tractable as any variable of natural science, and so really do bid fair to solve the problem of tastes, or rather to dissolve it. The biological treatment of preferences is one Becker certainly entertains. In a long footnote to the claim that there can be no disputes about tastes (previously quoted), he writes:

> To venture one further step, if genetical natural selection and rational behavior reinforce each other in producing speedier and more efficient responses to changes in the environment, perhaps the common preference function has evolved over time by natural selection and rational choice as that preference function best adapted to human society. That is, in the short run the preference function is fixed and households attempt to maximize the objective function subject to their resource and technology constraints.

> But in the very long run, perhaps those preferences sur-
> vive which are most suited to satisfaction given broad
> technological constraints on human society (e.g. physical
> size, mental ability, et cetera). (P. 145)

Note that Becker's claim is not that choices are genetically determined but that preferences are. Since preferences are not sufficient for choices, there is no tincture of "biological determinism" in this "speculation." Presumably a full specification of the
biological needs that the individual *Homo sapiens* is functionally
organized to meet will explain all or most of his behavior, including his deliberate choices. At present no such list is available, and even when it becomes available, it probably will not
divide up human behavior into parts that conform neatly to the
economist's object of inquiry. Even holding productive techniques constant, it will leave some intra- and interpersonal differences in behavior unexplained. These facts limit the degree
to which we can expect improvements in the systematic power
of a theory like Becker's, even as more is learned about human
needs and how they are fulfilled. In effect, the error term in any
specification of shared preferences is not likely to be eliminated,
and some rational choices must remain unaccountable. Nevertheless, what the study of human needs can provide should
certainly improve the schematic theory Becker offers us. Thus,
Becker's theory solves or rather dissolves the problem of tastes.
It does so by holding them constant, so that they drop out as
variables we need to cite in order to explain intra- and interpersonal differences. However, holding them constant will only
solve the problem if they turn out to be needs and not mere
wants.

When we turn to welfare economics, the treatment of Becker's stable preferences as needs seems to have even greater advantages, ones that are not reduced or qualified by our
temporary ignorance about all the needs an individual has. Or
so it seems. For Becker the application of the new theory to welfare economics is obvious:

> If observed differences in behavior as assumed to result
> from differences in tastes, and if the satisfaction of each
> person's tastes is used as a guide to normative statements,

then differences in behavior cannot be judged normatively. If, however, the observed behavior is assumed to result from different efficiencies with the same set of tastes, these can be judged by the level of full real income which they produce, i.e. by their level of productivity. For example, if education is said to alter tastes, one cannot speak of the effects of education on the level of utility: what is preferable to the college graduate may not be so to the grade school drop out and the two cannot, even in principle, reach agreement on which set of tastes is "better." But these judgements can be made if education affects the efficiencies of household production functions. Whatever yields greater commodity output is preferable and can be considered as such by both parties. (Pp. 145–46)

Of course, this conclusion will not follow unless the strong claims about the identity of preference functions and the characteristics of their components are right. Otherwise, there is still scope for exogenous differences to intrude and make the interpersonal comparison indeterminate. Moreover, the effect of education on productive efficiencies must be completely separate from its effect on preferences. Indeed, in Becker's theory, it can have no effect on preferences at all: still another reason to call them basic needs.

CAN WE "NATURALIZE" INFORMATION?

Let us assume, or at least hope, that economics, together with other disciplines, will ultimately derive a schedule of needs common enough among humanity to serve as the basis for calculating the amounts of the household commodities we all produce. Thus we may hope to make tastes endogenous, naturalizing them as the explicable expression of biologically accessible human needs. But consider the weight this approach throws on the economic theory of information.

With variation in beliefs and desires, it is possible to attribute differences in choice to either differences in beliefs or differ-

ences in desires. Under conditions of certainty, there is no variation in belief, so all differences in choice must be caused by
differences in desires. Of course, if we stipulate that these differences are exogenous and unaccountable, then microeconomic theory has almost nothing to tell us about individual
choice. If we attribute to each agent exactly the same preferences, then under conditions of certainty, every agent will
choose exactly the same bundle of commodities. But the most
obvious fact about people is that they make different purchases
in the market. We need, therefore, an account of behavior that
at least makes these differences possible.

Since this possibility of differences in preference among market goods is not accounted for by differences in preferences
with respect to household goods, the only explanation for these
differences must be found in differences among agents' techniques of production, described by E, Becker's environmental
variable, in equation (2) above. As noted above in the discussion
of the new theory's proclivity for paternalism, techniques of
production can best be understood as information acquired,
stored, and utilized by agents in the transformation of market
goods into household ones.

What Becker's theory does is throw the whole weight of explaining differences in choice among individuals, and eventually
all other shifts revealed by comparative statics, on differences
and changes in information acquired, stored, and utilized.
What the new theory needs is an account of belief in general,
which will explain how agents come to have different beliefs
about their own needs and available productive techniques, and
an account of the economics of information, as it affects markets, industries, and whole economies. An economic theory of
information will in fact become the entire substance of economics, since information will be the only causal variable at its
disposal. There is of course already a considerable body of work
in this area. Interestingly, much of it would be rendered obso-
lete by the adoption of Becker's approach, whereas other parts
of it would acquire increased importance.

Much of the economic theory of information is devoted to
the study of "incentive-compatible" arrangements, that is, mar-

ket mechanisms that give the self-interested agent an incentive to provide correct information about the agent's preferences.[7] It is well known that under anything approaching realistic conditions (i.e., a finite number of agents), no market mechanisms will in fact be incentive compatible, agents will be motivated to misinform others as to their preferences, and as a result Pareto-efficient allocations will be unattainable. However, if preferences are uniform, as Becker's theory would have it, then the problem of finding a market mechanism that induces agents to reveal them is dissolved. There is no need to find such a mechanism, because we already know what they are.

But of course, there is no free lunch. We have not really circumvented this serious economic problem, we have simply relabeled it. For now, instead of having inducements to misreport their preferences, they will have inducements to misreport their productive techniques. By doing so, agents can influence the supply of market goods, and thus their price, so that depending on which productive technique they employ they will secure a large quantity of one or another household good at lower prices than they would otherwise. Agents may misrepresent in at least two obvious ways: either they may claim to be employing more efficient techniques than they actually use, methods that require smaller amounts of market goods, thus lowering prices. Or they may have more efficient techniques than they reveal, which enable them to produce more of some household goods with the same amount of market goods. General knowledge of such a technique would result in increased resources to expend on other household goods and greater demand for the requisite market goods, thus raising their prices for all.

Initial studies devoted to the economics of information quickly settled on an agenda familiar from more traditional areas of economic research. If we treat techniques of production as inputs, can we prove the existence, uniqueness, and stability of a market-clearing equilibrium price vector? The answer given so

7. The explicit study of such issues begins with L. Hurwicz, "Optimality and Informational Efficiency in Resource Allocation Processes," in *Mathematical Methods in the Social Sciences,* edited by K. Arrow, S. Karlin, and P. Suppes (Stanford: Stanford University Press, 1960), pp. 27–46.

far is probably not, and the reason is in part because information about production is not used up as it is employed but may be transmitted to those who did not "pay" for it.

Surely, it will be agreed in general that information is a more tractable explanatory variable than taste or preference, at least in the sense that it is less "subjective." There is a noncontroversial difference between true and false, reliable and unreliable, relevant and irrelevant information, etc. By contrast, there seems to be no agreement on whether preferences can be distinguished into any cognitively interesting categories (*De gustibus non est desputandum*). In the long run and on average the employment of productive techniques based on false information will decline, as negative feedback discourages the employment of such techniques and encourages switching to techniques based on correct information. In making such assumptions, economic theory need not even suppose that the vehicle for the storage and utilization of information is the individual belief or expectation of the traditional theory. Naturally, information will have to affect individual choice behavior, but exactly how it does so—via consciously accessible beliefs, straightforward imitation of others, habit formation, etc.—is something on which the theory can be agnostic.

It will in fact have to be agnostic about the effect of information on individual choice: if economics takes Von Neumann–Morgenstern methods seriously, taking individual expectations seriously requires us to take preferences seriously, and this is what we want to avoid. Note that we cannot use an agent's schedule of needs to get the agent to reveal expectations about what productive techniques to employ, because the agent's expectations about this matter will hinge on expectations about what market goods are available at what prices. In other words, we must already assume the agent has information in order to employ nonintentional facts about him or her, like the agent's needs, in order to get at his or her expectations. The holism of the mental that we traced in chapter 5 makes the individuation of informational states as problematical as ever, even when we have "solved" the problem of individuating preferences.

In effect, we end up treating the economic agent as a black box, through which information predictably affects economic choice, without having to decide exactly how it does so. All we need to say is that the individual combines information with a schedule of needs or stable preferences to choose market goods and this influences economically interesting outcomes. In fact, the intractability of individual expectations can provide the economist with a new and better argument for the ancient claim that economic theory does or should concern itself only with the workings of aggregates, like markets, industries, or whole economies, and that claims about the individual are but convenient stepping stones to such concerns or heuristic devices for expressing findings about these aggregates. Current theory assumes that economically significant facts are uncovered and employed with varying degrees of efficiency, either by all individuals in a market or by some arbitrageurs. It is when they are uncovered that information is created, and new expectations are formed. Contemporary theorists devote considerable attention to whether information will be sought, transmitted, and exchanged for other commodities at optimal levels. By and large the answer seems to be no. But if there is in principle no way of establishing whether, when, and exactly which piece of information is in fact transmitted from one individual to others, because of the holistic character of the informational store of individual economic agents, then there is no obligation on the part of economics to trace out these chains. Instead, it should simply concentrate on the correlations between economically significant facts and the traditional explananda of the theory: changes in price, in demand, in supply, and so on, without worrying about underlying mechanisms that are in principle inaccessible.

If economists choose this course of agnosticism about the way information is linked to outcomes of economic interest, then they pretty well reconcile themselves to generic prediction. Without further details about the individuals through which the information affects the markets, economic theory will never be able to say exactly when or exactly how much information affects markets. Assume that once a new piece of information is fed into a market, the market moves to a new equilibrium reflecting the assimilation of this new information. If we want to know how far this equilibrium is from the previous one, or how long it takes to get there, we need to know more. We need to know how the information diffuses throughout the economy, among which agents first, second, and third; what they pay for the information, if anything (this will affect the new equilibrium in other commodities); what individuals *believe* the relevance of this information is for economic decisions they make, etc. Of course, we can explore the impact of a piece of information on a representative, average, or typical agent. But in doing so, we are simply treating this agent as a convenient substitute for the market as a whole, and our results will continue to be at most generic. We can improve our predictions and our explanations by stratifying the market into agents with different assets and different decisions to make, who may secure the information at different times, but such an approach begins to give up the desired neutrality about what goes on in the heads of individual economic agents.

The role of uncertainty—the lack of information—in economic theory reflects this limitation to generic theory. It is only by introducing uncertainty that economics can generate the most obvious features of a real economy: money, futures markets, shares and stock markets, bonds and interest rates, insurance, etc. The theoretical strategy is to assume uncertainty about future or present prices, to derive rational agents' responses to these uncertainties, and then to shape these derivations into forms that will enable us to prove the existence of equilibrium outcomes. What we need in order to say more— that money will be demanded or that the interest rate will be nonnegative—are hypotheses about how individuals respond to these uncertainties, that is, about their behavior in the light of their expectations. However, then we need to measure expectations more precisely than we can.

By refusing to take seriously the psychological states it attributes to individuals, economics ends up treating the economic agent as just a complicated and highly discriminant teleological system, one whose behavior is highly appropriate to its biological needs, given its environment. The economic agent is an organism whose sensory apparatus is sufficiently sensitive to track most changes in the informational stimuli its environment provides, including information about the best ways to meet its needs, and whose memory capacity is large enough and reliable enough to store information for future use.

The difference between this minimal teleological conception of the agent and rational choice theory is largely one of specificity: rational choice theory imputes a specific mechanism to the economic agent. It is a mechanism that not only realizes the generally teleological processes we actually evince but goes on to an even finer-grained teleology. The assumption of rational choice makes us out to be even more efficient in information detection, computation, etc. It attributes to us a mechanism for doing this that we cannot realize, because we are not really creatures whose behavior is the result of omniscient constrained maximization of complete and transitive preferences.

A minimal teleological theory's chief advantage over the conventional assumptions of microeconomics are its agnosticism about the procedural rationality of individual agents. Its implications about their substantive rationality, or at least about the aggregation of their choices into market demand and supply curves, will be no different from those of the standard theory. And since this nonintentional approach makes no claim about exactly how, when, or whether specific changes in information have specific effects on the mechanisms that issue in choices, this approach holds out no more hope of transcending generic predictions than the old theory. For our purposes, it is an improvement in form and not in content.

Moreover, just as revealed preference theory simply eliminates by fiat an important range of activity that economists have traditionally concerned themselves with, refusing to take information seriously as an intentional state would also force economics to turn its back on an area of great concern to important theorists. Beginning at least with the work of Hurwicz, economists have explicitly concerned themselves with the efficiency effects of decentralizing information among agents and among organizations of agents. The models developed make quite explicit assumptions about the beliefs of agents about their own environments and those of other agents. They go on to derive allocations in an economy as a consequence of these expectations and the communications that they lead agents to engage in. These models attribute explicit languages, messages, responses, memory, and other properties to agents incompatible with a view of economics that makes the individual a mere *façon de parler* for aggregates. As one would expect, one central question explored by such models is whether the resulting allocations are optimal equilibria. If the problems of intentionality canvassed in the last chapter are as grave as claimed, it is hard to see how the elegant theorems derived in these models can be parlayed into a predictively improvable theory.

GENERIC PREDICTIONS AND THE TEMPTATION OF BIOLOGY

I have argued that Becker's new theory of consumer behavior has much to recommend itself, provided that it is treated as a claim about biological needs. Becker himself has indicated the affinities of his approach to biological and, in particular, evolutionary thinking. He is not the first economist to seek support for neoclassical theory or its variants in actual or potential affinities to the theory of natural selection. To some economists drawing parallels between economic theory and evolutionary theory constitutes part of a powerful defense of the empirical status of economic theory. The analogy to evolutionary biology is a particular favorite among economists (unlike Becker) who decline to take seriously the actions of individual agents as an explanatory subject for economics. They argue that just as evolutionary theory focuses on populations, economic theory need make no claims about nor be tested by implications for individual choices. Thus, these economists hold that the fact that we cannot eliminate the intentionality of information is no obstacle to predictive success where we really seek it: in market demand and supply, for instance. In the long run individual differences in beliefs about economic variables must average into aggregate expectations discounted by efficient markets that track the facts about the economy, just as individual biological variation aggregates into changing population adaptations that track the environment. What is sauce for the Darwinian goose is sauce for the economist's gander.

However, as we shall see, Darwinian theory is a remarkably inappropriate model for economic theory. The theory of natu-

177

ral selection shares few of its strengths and most of its weaknesses with neoclassical theory. "Biologizing" economic theory much beyond Becker's innovation turns out to be the most powerful argument for the claim that neoclassical economics is fated to at most generic predictions. For unbeknownst to most economists who succumb to this temptation, the theory of natural selection is itself limited largely to generic predictions.[8]

The theory of natural selection is breathtaking in its simplicity. Darwin began with some obvious observations. The first was that notwithstanding the Malthusian assertion that organisms reproduce geometrically, the population of most species remains constant over time. From this it follows that there is a struggle for survival, both within species and between members of differing species. Darwin's second observation was that species are characterized by variation among their members. Darwin inferred the survival of those variants that are most fitted to their environments—most able to defend themselves against predators, find shelter against the elements, provide themselves with food, and therefore most able to reproduce in higher numbers. If these traits are hereditary, then they will be represented in higher proportions in each generation until they become ubiquitous throughout the species. This will be especially true for hereditary traits that enhance an organism's ability to procure mates and otherwise ensure the reproduction of fertile offspring. It is crucial to Darwin's theory that variation is large and blind—in any generation there will be differences on which selection can operate, and these differences are not elicited by environmental needs but are randomly generated. Darwin knew nothing of genetics, and his theory requires only that there be variation and heredity. Modern genetic theory provides for both of these requisites of Darwin's theory. For this reason genetics is often treated as part of evolutionary theory. Selection, for Darwin, is a misleading metaphor, since Darwin's theory deprives nature of any purpose, teleology, design, or intentions of the sort the notion of selection suggests. Nature se-

8. Oddly enough, McCloskey is something of an exception to this claim. But McCloskey mistakenly holds that evolutionary theory is without any predictive power. See chap. 2, the second section.

lects only in the sense that the match or mismatch between the environment and the fortuitously generated variants determines survival and thus reproduction.

So much for the bare bones of the theory—now some of the details. Darwin knew little about the sources of variation. We know more. Some variation is produced by mutations, but not enough to account for the diversity and the adaptations we actually observe. Most variation, especially among sexually reproducing organisms, is the result of the shuffling of genes through interbreeding within a species. The interaction of different genes with one another and with various features of an environment produces a range of phenotypes which are selected for in accordance with their strictly fortuitous contributions to or withdrawals from survival and reproduction.

However, a single new variant (produced by recombination or mutation) no matter how adaptive is likely to be swamped in its effects if it appears in a large population. One long-necked giraffe in a million is just not going to make a difference. To begin with, though she can reach food other animals cannot, she just might be hit by lightning and die before breeding. For another, when the one copy of the gene for long necks combines with any of the million or so copies of the gene for short necks, the result may just be short necks. For the long-neck gene to make a difference, the number of giraffes with which its bearer breeds must be small, so that copies of the gene in subsequent generations have a chance to combine with one another and produce more long necks. Thus the structure of the evolving population is important for the occurrence of adaptive evolution: it should be small in size and allow a certain amount of interbreeding. If it is too small, however, a well-adapted population could be wiped out by random forces before it has a chance to expand its numbers.

In addition, for adaptive evolution the environment must remain relatively constant over long periods of time. The environment presents organisms with survival problems, but it takes a generation for the best solutions to these problems to make a difference for the species. For it is only in the proportions of the offspring that the best solutions to an adaptational problem have their evolutionary effects. If an environment changes faster than the generation time of a species, the species will never show any pattern of adaptation. Among animals this is not a serious problem. Most environmental challenges—heat, cold, gravity, lack of food in winter, etc.—have been with us for literally geological epochs. And even the generation time of the tortoise—a hundred and fifty years or so—is as nothing compared to such epochs. So, there has been time and enough stability for a lot of evolution. Nevertheless, critics of Darwinian theory complain that although there has been enough time for the evolution of species, there seems little evidence of transitional forms of the sort we should expect. Indeed, it is an old saw among paleontologists that the fossil record shows mainly that evolution took place elsewhere. Another important thing to note is that in rapidly changing environments, survival favors generalists who are moderately well adapted to a number of environments over specialists who are very well adapted to just one sort of environment.

In selecting variants for differential survival, nature works with what variation presents and shapes available properties. In effect, it seeks the quick and dirty solution to an adaptive problem and not the optimally adaptive one. Because an organism can make no contribution to evolution unless it survives, nature will work with what is presented to it and will encourage early approximate solutions over late but elegant and exact ones. By the time an elegant solution is available, the lineage may be extinct.

Another important feature of Darwinian evolution is its commitment to individual selection. Darwin shares with modern biologists a conviction that the locus of selection is the individual organism, and not larger groups in which individuals participate. If groups of various kinds evolve, then it will be because of the adaptational advantages they accord to individuals who maximize fitness. Groups which disadvantage some of their individual members in order to increase average fitness, for example, are vulnerable to free riders who take advantage of benefits groups provide their members while failing to contribute to group welfare. These free riders will prosper at the expense of contributors until they have completely displaced them. For groups without enforcement mechanisms, individual selection will always swamp group selection. Moreover, it is hard to see how enforcement mechanisms can emerge in the first place, given individual fitness maximization. In this respect the theory of natural selection, like microeconomics, honors the principle of methodological individualism.

Finally, there is another feature of evolutionary theory well worth understanding: its relatively weak powers of prediction. About the only place where there is very strong predictive evidence for natural selection is in laboratory experiments and in what animal and plant breeders call artificial selection. In the laboratory and on the farm, we can control environmental conditions (reproductive opportunities) stringently enough that only a narrow class of animals survive and reproduce. The result is relatively rapid changes in the proportions of properties adaptive to our interests as breeders. But not only have we not produced anything that all will agree constitutes a new species, but as noted above, the fossil record does not help either. Evolutionary biology has no striking retrodiction to its credit, and such predictions as it might make are either very generic or likely to be no more reliable than the initial or boundary conditions to which the theory is applied.

In fact, for much of the century the theory of natural selection has been stigmatized as completely lacking in evidential bearing, as being an unfalsifiable trivial tautology. The charge is well understood. The theory asserts that the fittest survive and reproduce differentially. But the only applicable uniform quantitative measure of fitness is reproductive rates. Accordingly the formula becomes: those with the highest reproductive rates have the highest reproductive rates. It is therefore no wonder that no evidence can be found which contradicts the theory, nor can we expect to find evidence that strikingly confirms it either. It is no good defending evolutionary theory by rejecting the demand that theory be falsifiable. To do so is just to blame the messenger. Even if strict falsifiability is too stern a test for a scientific theory, it's still a serious weakness of any theory if it cannot enable us to identify its causal variables independently of the effects they bring about. And this is indeed the problem for evolutionary theory, just as it is for economic theory. A better response to this complaint against the theory is to admit that in general we cannot enumerate what fitness consists in—there are too many determinants of evolutionary fitness to be mentioned in a theory—whether and how much a property conduces to fitness depends on the environment, and the only thing all determinants of fitness differences have in common is their *effect* on rates of reproduction. So it is natural to measure fitness differences in terms of their common effects. Once we are clear on the difference between fitness and what we use to measure it, the claim that the fittest survive and reproduce in higher numbers is no more vacuous than the claim that increases in heat make thermometers rise. This insight rebuts the charge that evolutionary theory is vacuous or tautologous, but does nothing to alleviate its predictive weakness.[9]

There are many ways in which organisms can adapt in response to a given environmental constraint. An ice age presents survival problems that can be solved by growing fur, adding layers of fat, changing shape to minimize surface area, migration, or hibernation. There are many ways in which an environment can change: temperature, humidity, wind, pressure, flora, fauna, CO_2 concentration, for example. Multiplying the environmental changes times the number of different adaptational responses to each change makes it clear that interesting generalizations about adaptation are not to be found in the expression of the theory itself. In fact, because of this, the formalism of the theory has pretty much taken the form of stochastic models of changes in gene frequencies. Making certain assumptions about the independence of genes (and therefore observable traits of organisms—phenotypes) from one another and adding assumptions about differences in fitness, size of interbreeding population, and so on, the biologist can derive conclusions about the change in gene frequencies over time. The question then becomes whether there are biological phenomena that realize the assumptions of the model well enough that its consequences can guide our expectations about the phenom-

---

9. For a discussion of the predictive weakness of the theory of natural selection and the charge that it is a grand tautology, see Alexander Rosenberg, *The Structure of Biological Science* (Cambridge: Cambridge University Press, 1984), chaps. 5–7.

ena. Instead of seeking general laws about the way in which environmental changes result in adaptations, evolutionary biologists consider which models of changes in gene frequencies most clearly illuminate processes of interest, and whether the most illuminating models have interesting features in common. By and large the number of such interesting models has not been great, and they have had relatively few distinctive features in common. This should be no surprise. If the models were very successful and had a good deal of structure and a large proportion of assumptions in common, then the most obvious explanation of these facts would be the truth of a simple and powerful theory that unified them all and explained specifically why they worked so well. Such a theory would in fact replace the theory of natural selection, whose weakness and lack of predictive content lead biologists to seek models of restricted case studies instead of predictively powerful general laws.

DARWIN, FRIEDMAN, AND ALCHAIN

Why should anyone suppose Darwinian evolutionary theory can provide a useful model for how to proceed in economics? One apparently attractive feature of the theory for economists is the methodological defense it seems to provide neoclassical theory in the face of charges that the latter theory fails to account for the actual behavior of consumers and producers. Thus, Friedman offers the following argument for the hypothesis that economic agents maximize money returns:

> Let the apparent immediate determinant of business behavior be anything at all, habitual reaction, random chance, or whatnot. Whenever this determinant happens to lead to behavior consistent with rational and informed maximization of returns, the business will prosper, and acquire resources with which to expand; whenever it does not, the business will tend to lose resources and can be kept in existence only by the addition of resources from outside. The process of "natural selection" thus helps to validate the hypothesis or, rather, given natural selection, accep-

tance of the hypothesis can be based on the judgement that it summarizes appropriately the conditions for survival.[10]

This argument does reflect a feature of evolutionary theorizing, though admittedly a controversial one. The natural environment sets adaptational problems that organisms have to solve in order to survive. The fact that a particular species is not extinct is good evidence that it has solved some of the problems imposed upon it. This fact about adaptational problems and their solutions plays two roles in evolutionary thinking. First, examining the environment, biologists might try to identify the adaptational problems that organisms face. Second, focusing on the organism, they sometimes attempt to identify possible problems which known features of the organism might be solutions to. The problem with this approach is the temptation of Panglossianism: imagining a problem to be solved for every feature of an organism we detect. Thus, Dr. Pangloss held that the bridge of the nose was a solution to the adaptational problem of holding up glasses. The problem with inferences from the environment to adaptational problems is that we need to determine all or most of the problems to be solved, for each of them is an important constraint on what will count as solutions to others. Thus having a dark color will not be a solution to the problem of hiding from nocturnal predators unless the organism can deal with the heat that such color will absorb during the day. On the other hand, a color that will effect the optimal compromise between these two constraints may fail a third one, say, being detectable by conspecifics during mating season.

Then there is the problem of there being more than one way to skin a cat. Even if we can identify an adaptational problem and most of the constraints against which a solution can be found, it is unlikely that we will be able to narrow the range of equally adaptive solutions down to just the one that animals actually evince. Thus, we are left with the explanatory question of why this way of skinning the adaptational cat emerges and not another apparently equally as good a one. There are two answers to this question. One is to say that if we knew all the con-

10. "The Methodology of Positive Economics," in *Essays in Positive Economics* (Chicago: University of Chicago Press, 1953), p. 35.

straints, we would see that the only possible solution is the actual one. The other is to say that there is more than one equally adequate solution, and that the one finally "chosen" appeared for nonevolutionary causes. The first of these two replies is simply a pious hope that more inquiry will vindicate the theory. The second in effect limits evolutionary theory's explanatory power and denies it predictive power.

These problems have in general hobbled "optimality" analysis as an explanatory strategy in evolutionary biology. Many biologists find the temptations of Panglossianism combined with the daunting multiplicity of constraints on solutions to be so great that they despair of providing an evolutionary theory that contributes to our detailed understanding of organisms in their environments.

The same problems bedevil Friedman's conception and limit the force of his conclusion. The idea that rational informed maximization of returns sets a necessary and/or sufficient condition for long-term survival in every possible economic environment, or even in any actual one, is either false or vacuous. Is the hypothesis that returns are maximized over the short run, the long run, the fiscal year, the quarter? If we make the hypothesis specific enough to test it is plainly false. Leave it vague, and the hypothesis is hard to test. Suppose we equate the maximization of returns hypothesis with the survival of the fittest hypothesis. Then nothing in particular follows about what economic agents do and how large their returns are, any more than it follows what particular organisms do and how many offspring they have. However many offspring, however great the returns, the results will be maximal, given the circumstances, over the long run. What we want to know is what features of organisms increase their fitness, what strategies of economic agents increase their returns. And we want this information both to explain particular events in the past and to predict the course of future evolution. For the hypothesis of maximization of returns to play this substantive role, it cannot be supposed to be on a par with the maximization of fitness hypothesis. Rather, we need to treat it as a specific optimal response to a particular environmental problem, rather like we might treat coat color as an optimal response to an environmental problem of finding a color that protects against predators, does not absorb too much heat, is visible to conspecifics, etc. But when we think of the maximization of returns hypothesis this way, it is clear that maximizing dollar returns is not a condition of survival in general, either in the long run or the short run.

As noted above, nature has a preference for quick and dirty solutions to environmental problems. It seems to "satisfice," in Herbert Simon's phrase. But unlike satisficing, nature's strategy really may be a maximizing one. It may be that the constraints are so complicated and so unknown to us that the solutions selection favors look quick and dirty to us. If we knew the constraints we would see that the solutions are elegant and just on time. Learning what the constraints are and how the problems are solved is where the action is in vindicating the theory of natural selection, because only that will enable us to tell whether the solution really maximizes fitness, as measured by offspring. Similarly, in economics the action is in learning the constraints and seeing what solutions are chosen. Only that will tell us whether dollar returns are really maximized and whether maximizing dollar returns ensures survival. To stop where Friedman does is to condemn the theory he sets out to vindicate to the vacuity with which Darwinian theory is often charged.

If the theory of natural selection is to vindicate economic theory or illuminate economic processes it will have to do more than just provide a Panglossian assurance that whatever survives in the long run is fittest. What is needed in any attempt to accomplish this is a better understanding of the theory of natural selection. Such an improved understanding of the theory is evident in Alchain's approach to modeling economic processes as evolutionary ones.[11]

Alchain's approach does not give rise to obvious Panglossian objections, nor does it make claims about empirical content which transcend the power of an evolutionary theory to deliver. Still, its problems reveal more deeply the difficulties of taking an evolutionary approach to economic behavior.

11. A. Alchain, "Uncertainty, Evolution and Economic Theory," *Journal of Political Economy* 58 (1950): 211–21. Page references in this section are to this paper.

To begin with, Alchain's approach acknowledges that Darwinian theory's claims about individual responses to the environment are hard to establish, impossible to generalize, and therefore without predictive value for other organisms in other environments. Alchain recognized that the really useful versions of evolutionary theory are those which focus on populations large enough that statistical regularities in responses to environmental changes can be discerned. And he recognized that Darwinian evolution operates through solutions to adaptational problems that are in appearance at any rate quick and dirty, approximate and heuristic, not rationally and informationally maximizing. Like the biological environment, the economic one need not elicit anything like the maximization of returns that conventional theory requires:

> In an economic system the realization of profits is the criterion according to which successful and surviving firms are selected. This decision criterion is applied by an impersonal market system . . . and may be completely independent of the decision processes of individual units, of the variety of inconsistent motives and abilities and even of the individual's awareness of the criterion.
>
> The pertinent requirement—positive profits through relative efficiency—is weaker than "maximized profits," with which, unfortunately, it has been confused. Positive profits accrue to those who are better than their actual competitors, not some hypothetically perfect competitors. As in a race, the award goes to the relatively fastest, even if all competitors loaf.
>
> . . . success (survival) accompanies relative superiority; . . . it may . . . be the result of fortuitous circumstances. Among all competitors those whose particular conditions happen to be the most appropriate of those offered to the economic system for testing and adoption will be "selected" as survivors. (Pp. 213–14)

Alchain also recognizes that adaptation is not immediate and is only discernable to the observer in the change in statistical distributions over periods of time and that what counts as adaptive will change as the economic environment does. Alchain uses a parable to illustrate the way that the economic environment shifts the distribution of actually employed choice strategies toward the more rational:

> Assume that thousands of travellers set out from Chicago, selecting their roads completely at random and without foresight. Only our "economist" knows that on but one road there are any gas stations. He can state categorically that travellers will continue to travel only on that road: those on other roads will soon run out of gas. Even though each one selected his route at random, we might have called those travellers who were so fortunate as to have picked the right road wise, efficient, farsighted, etc. Of course we would consider them the lucky ones. If gasoline supplies were now moved to a new road, some formerly luckless travellers again would be able to move; and a new pattern of travel would be observed, although none of the players changed his particular path. The really possible paths have changed with the changing environment. All that is needed is a set of varied, risk-taking (adaptable) travellers. The correct direction of travel will be established. As circumstances (economic environments) change, the analyst (economist) can select the type of participants (firms) that will now become successful; he may also be able to diagnose the conditions most conducive to greater probability of survival. (P. 214)

To ensure survival and significant shifts in the direction of adaptation, several other conditions must be satisfied. To begin with, the environment must remain constant long enough that those strategies more well adapted to it than others will have time to outcompete the less well adapted and to increase their frequency significantly enough to be noticed. Moreover, the initial relative frequency of the most well adapted strategy must be high enough that it will not be stamped out by random forces before it has amassed a sufficient advantage to begin to displace competitors. And of course it must be the case that there are significant differences among competing strategies. Otherwise, their proportions at the outset of competition will remain constant over time. There will be no significant changes in proportions to report.

What kind of knowledge will such an economic theory provide? Even at his most optimistic Alchain was properly limited in his expectations. He made no claims that with an evolutionary approach the course of behavior of the individual economic agent could be predicted. Here the parallel with evolution is obvious. Darwin's theory not only has no implications for what will happen to any individual organism, its implications for large numbers of them are at best probabilistic:

A chance dominated model does not mean that an economist cannot predict or explain or diagnose. With a knowledge of the economy's requisites for survival and by a comparison of alternative conditions, he can state what types of firms or behavior relative to other possible types will be more viable, even though the firms themselves may not know the conditions or even try to achieve them by readjusting to the changed conditions. It is sufficient if all firms are slightly different so that in the new environmental situation those who have their fixed internal conditions closer to the new, but unknown optimum position now have a greater probability of survival and growth. They will grow relative to other firms and become the prevailing type, since survival conditions may push the observed characteristics of the set of survivors towards the unknowable [to them] optimum by either (1) repeated trials or (2) survival of more of those who happen to be near the optimum—determined ex post. If these new conditions last "very long," the dominant firms will be different ones from those which prevailed or would have prevailed under the other conditions. *Even if* the environmental conditions cannot be forecast, the economist can compare for given alternative potential situations the types of behavior that would have higher probability of viability or adoption. If explanation of past results rather than prediction is the task, the economist can diagnose the particular attributes which were critical in facilitating survival, even though individual participants were not aware of them. (Emphasis added, p. 216)

As a set of conditional claims, most of what Alchain says about the explanatory and predictive powers of an evolutionary theory of economic processes is true enough. The trouble is that almost none of the conditions obtain, either in evolutionary biology or in economic behavior, that would make either theory as useful as Alchain or any economist needs it to be. Thus, the attractions of an evolutionary theory for economists must be very limited indeed. Alchain rightly treats the economy as the environment to which individual economic agents are differentially adapted. As with the biological case, we need to know what "the requisites of survival" in the environment are. In the biological case this is not a trivial matter, and beyond the most obvious adaptational problems, there are precious few generalizations about what any particular ecological environment requires for survival, still less what it rewards in increased reproductive opportunities. We know animals need to eat, breathe, and avoid illnesses and environmental hazards, and the more of their needs they fulfill the better off they are. But we do not know what in any given environment the optimal available diet is or what the environmental hazards are for each of the creatures that inhabit the environment. And outside ecology and ethology, few biologists are interested in this information in any case, for its systematic value to biology is very limited. Ignorance about these requisites for survival in biology make it difficult to predict even "the types of . . . behavior relative to other possible types [that] will be more viable." It is easy to predict that all surviving types will have to subsist in an oxygen-rich environment, where the gravitational constant is 32 feet/second^2, and the ambient temperature ranges from 45 degrees to -20 degrees Celsius. But such a "prediction" leaves us little closer to what we hope to learn from a prediction. The same must be true in evolutionary economics. We have no idea of what the requisites for survival are, and even if we learned them, they would probably not narrowly enough restrict the types that can survive to enable us to frame any very useful expectations for the future. Of course, this is not an in principle objection to an evolutionary approach. But consider what sort of

information would be required to establish a very full list of con-
crete necessary conditions for survival of, say, a firm in any very
specific market, and then consider the myriad ways in which economic agents could act to satisfy those conditions. This information is either impossible to obtain or else if we had it, an
evolutionary approach to economic processes would be superfluous. To see this, go back to Alchain's discussion.

Alchain notes that over time the proportion of firms of various types should change: the proportion of those that are fitter
should increase while those less fit should decrease. If environmental conditions last a long time, "the dominant firms will be
different from those which prevailed . . . under the other conditions." True enough, but what counts as environmental conditions lasting a long time? In the evolutionary context "long
enough" means at least one generation, and the duration of a
generation will vary with the species. In addition the notion of
"long enough" reflects a circularity which haunts evolutionary
biology. Evolution occurs if the environment remains constant
long enough for the proportion of types to change. "Long
enough" is enough time for the proportions to change. Moreover, when the numbers of competing individuals are small,
there may be a change in proportions of types that is not adaptational but is identified as "drift"—a sort of sampling error.
But what is a small number of individuals versus a large number? Here the same ambiguity emerges. "Large enough" means
a number in which changes in proportion reflect evolutionary
adaptation. The only way to break out of this circularity of
"long enough, large enough," etc., is to focus on individual
populations in particular environments over several generations. And the answer we get for any one set of individuals will
be of little value when we turn to another set of the same types
in different environments or different types in the same environment.

Can the situation be any better in economics? In fact, won't
the situation be far less promising? After all, the environment
within which an economic agent must operate does not change
with the stately pace of a geological epoch. Economic environments seem to change from day to day. If they do, then there is
never enough time for the type most adapted to one environ-

ment to increase its proportions relative to other types. Before it has had a chance to do so, the environment has changed, and another type becomes most adapted. But perhaps economic environments do not change quite so quickly. Perhaps to suppose that they do change so quickly is to mistake the weather for the climate. Day-to-day fluctuations may be a feature of a more long-standing environment. The individual most well adapted to an environment is not one that responds best to each feature of it, including its variable features, but that adapts best overall on an average weighted by the frequency with which certain conditions in the environment obtain. So, the period of time relevant to evolutionary adaptation might be long enough for changes in proportion to show up. For the parallel to evolution to obtain, this period of environmental constancy will have to be longer than some equivalent to the generation time in biological evolution. But is there any such equivalent among economic agents? Is there a natural division among economic agents into generations? With firms the generation time might be the period from incorporation to the emergence of other firms employing the same method in the same markets through conscious imitation; with individual agents the minimal period for evolutionary adaptation might be the time it takes an individual to train another one to behave in the same way under similar economic circumstances. But these two periods are clearly ones during which the economic environment almost always changes enough to shift the adaptational strategy.

The only way we can use an evolutionary theory to predict the direction of adaptation is by being able to identify the relevant environment that remains constant enough to force adaptational change in proportions of firms. As Alchain tacitly admits, this is something we cannot do. He notes, "Even if the environmental conditions cannot be forecast, the economist can compare for given alternative potential situations the types of behavior that would have a higher probability of viability or adoption" (p. 217). This is a retrospective second best. Suppose economically relevant environmental conditions could be forecast. Then, it is pretty clear we would not need an evolutionary theory of economic behavior. Friedman's rationale for neoclassical theory would then come into its own. If we knew environmental conditions, then we could state what optimal adaptation to them would be. And if we could do this, so could at least some of the economic agents themselves. To the extent that they could pass on this information to their successors, Panglossianism would eventually be vindicated in economic evolutionary theory. Economic agents would conform their actions to the strategy calculated to be maximally adaptive, just as Friedman claims. An evolutionary theory of economic behavior is offered either as an alternative to rational maximizing or as an explanation of its adequacy. If rational maximizing is adequate as a theory, evolutionary rationales are superfluous; if it is not adequate, then an evolutionary approach is unlikely to be much better, and for much the same reason: neither economic theorists nor economic agents can know enough about the economic environment for the former's predictions or the latter's decisions to be regularly vindicated.[12]

EQUILIBRIUM AND INFORMATION IN EVOLUTION AND ECONOMICS

The next chapter is devoted to an examination of the fascination of economics with equilibrium thinking, but no discussion of the prospective parallels between economics and biology would be complete without a treatment of their shared fondness for equilibrium approaches. In fact, one of the features of evolutionary theory that makes it attractive to the economist is the role of equilibrium in the claims the theory makes about nature. Equilibrium is important for economic theory not least because of the predictive power it accords the economist. An economic system in equilibrium or moving toward one is a system some or all of whose future states are predictable by the

12. In this section I have focused on two classical statements of the evolutionary model for vindicating neoclassical economic theory. There is, however, another approach, developed in great detail by Richard Nelson and Sidney Winter in the *Evolutionary Theory of the Firm* (Cambridge: Belknap Press of Harvard University Press, 1984), in which evolutionary theory provides the model for a theory of economic behavior quite different from the conventional approach. For reasons I advance elsewhere, I fear that this approach offers no more predictive power than the conventional theory.

economist. Equilibrium has other (welfare-theory-relevant) aspects, but its attractions for economists must in part consist in the role it plays so successfully in physical theory and evolutionary theory. Evolutionary biology defines an equilibrium as one in which gene ratios do not change from generation to generation, and it stipulates several conditions that must obtain for equilibrium: a large population mating at random, without immigration, emigration, or mutation, and of course without environmental change. Departures from these conditions will cause changes in gene frequencies within a population. But over the long run the changes will move in the direction of better adaptation to the environment—either better adaptation to an unchanged one or adaptation to a new one. The parallel to economic equilibrium is so obvious that mathematical biologists have simply taken over the economist's conditions for the existence and stability of equilibria. If a unique stable market-clearing equilibrium exists, then its individual members are optimally adapted to their environment, no trading will occur, and there will be no change—no evolution—in the economy. But if one or another of the conditions for equilibrium are violated, an efficient economic system will move back to the original equilibrium or to a new one by means of adjustments in which individuals move along paths of increased adaptation.

In evolutionary biology equilibrium has an important explanatory role. So far as we can see, populations remain fairly constant over time, and among populations the proportions of varying phenotypes remain constant as well. Moreover, when one or another of the conditions presupposed by equilibrium of gene frequencies is violated, the result is either compensating movement back toward the original distribution of gene frequencies or movement toward a new distribution of gene frequencies. These facts about the stability of gene frequencies and their trajectories need to be explained, and the equilibrium assumptions of transmission genetics are the best explanations available. In addition, they will help us make *generic* predictions that when one or another condition, like the absence of mutation, is violated, a new equilibrium will be sought. Although we can sometimes even predict the direction of that new equilibrium, in real ecological contexts (as opposed to simple text-

194

book models) we can hardly ever predict that actual value of the new equilibrium distribution of gene frequencies. This is because we do not know all the environmental factors that work with a change in one of these conditions, and we have only primitive means to measure the dimensions of those factors that we do know about.

Now compare economics. To begin with, we have nothing comparable to the observed stability of gene frequencies that needs to be explained. So the principal explanatory motivation for equilibrium explanations is absent. We cannot even appeal to the stability of prices as a fact for equilibrium to explain in economics, because we know only too well that neoclassical general equilibrium theory has no explanation for price stability. That is, given an equilibrium distribution and a change in price, there is no proof that the economy will move to a new general equilibrium. (For this reason general equilibrium theory has recourse to the Walrasian auctioneer and *tâtonnement*.)

There is no doubt that economic equilibrium theory has many attractive theoretical features—mathematical tractability, the two welfare theorems—but it lacks the most important feature that justifies the same kind of thinking in evolutionary biology: independent evidence that there is a stable equilibrium to be explained.

One of the factors that gives us some confidence that equilibrium obtains with some frequency in nature is that changes in gene frequencies are not self-reinforcing. If something changes that has the effect of changing gene frequencies, then hardly ever will the change in gene frequencies precipitate still another round of changes in gene frequencies and so on, thus cascading into a period of instability. Of course, sometimes evolutionary change is "frequency dependent": if one species of butterfly increases in population size because it looks like another species that birds avoid, then once it has grown larger in number than the bad-tasting butterflies, its similar appearance and the genes that code for appearance will no longer be adaptive and may decline. But presumably, the proportions will return to some optimal level and be held there by the twin forces of adaptation and maladaptation.

In the game theorist's lingo, evolutionary adaptational problems are parametric: the adaptiveness of an organism's behavior does not depend on what other organisms do. But we cannot expect this absence of feedback in economic evolution. Among economic agents the problem is strategic. Economic agents are far more salient features of one another's environment than animals are features of one another's biological environment. Changes in agents' behaviors affect their environments regularly, because they call forth changes in the behavior of other agents, and these further changes cause a second round of changes in their behavior. Game theorists have come to identify this phenomenon under the rubric of the common knowledge problem. Economists traditionally circumvented this problem by two assumptions that have parallels in evolutionary biology as well. It is important to see that the parallels do not provide much ground for the rationalization of economic theorizing in the biologist's practice.

Both evolutionary equilibrium and economic general equilibrium require an infinite number of individuals. In the case of evolution this is to prevent drift or sampling error from moving gene frequencies independent of environmental changes. In the case of the theory of pure competition it is to prevent agent choice from becoming strategic. If the firm is always a price taker and can have no effect on the market, then it can treat its choices as parametric. Where numbers of interactors are small, the assumption of price taking produces badly wrong predictions, and there is indeed no stability and typically no equilibrium.

Is sauce for the biological goose sauce for the economic gander? Can both make the same false assumption with equal impunity? The fact is, though assumptions of infinite population size are false with respect to interbreeding populations, it seems to do little harm in biology. That is, despite the strict falsity of the evolutionary assumption, populations seem to be large enough that the theory making these false assumptions can explain the evident facts of constancy and/or stability of gene frequencies. In the case of economics, there are no such evident facts, and one apparent reason seems to be the falsity of the assumption of an infinite number of economic agents.

The other assumption evolutionary theory and economic theory traditionally make is that the genes and the agents are "omniscient." Genes carry information in two senses. They carry instructions for the building and maintenance of proteins and for assemblages of them that meet the environment as phenotypes. They also indirectly carry information about which phenotypes are most adapted to the environment in which they find themselves. They do so through the intervention of selective forces that cull maladaptive phenotypes and thus the genes that code for their building blocks. And so long as the environment remains constant, the gene frequencies will eventually track every environmentally significant biologically possible adaptation and maladaptation. In this sense the genome is in the long run omniscient about the environment. There are two crucial qualifications here: first, the assumption of the constancy of the environment, something economic theory has little reason to help itself to; second, there is the "long run," another concept evolutionary theory shares with economic theory. Evolutionary biology has world enough and time for theories that explain and predict only in the long run—geological epochs are close enough to infinite not to matter for many purposes. But Keynes pointed out the problem for economics of theories that explain only the long run. An evolutionary economic theory committed to equilibrium is condemned at best to explain only the long run.

We know only too well the disequilibrating effects of non-omniscience, that is, how information obstructs the economy's arrival at or maintenance of an equilibrium. Indeed, the effects of differences in information on economic outcomes are so pervasive that we should not expect economic phenomena ever to reflect the kind of equilibrium evolutionary biological phenomena do. Arrow has summarized the impact of information on equilibrium models succinctly:

> If nothing else there are at least two salient characteristics
> of information which prevent it from being fully identified
> as one of the commodities represented in our abstract
> models of general equilibrium: (1) it is, by definition, indi-
> visible in its use; and (2) it is very difficult to appropriate.
> With regard to the first point, information about a method

of production, for example, is the same regardless of the scale of the output. Since the cost of information depends only on the item, not its use, it pays a large scale producer to acquire better information than a small scale producer. Thus, information creates economies of scale throughout the economy, and therefore, according to well-known principles, causes a departure from the competitive economy.

Information is inappropriable because an individual who has some can never lose it by transmitting it. It is frequently noted in connection with the economics of research and development that information acquired by research at great cost may be transmitted much more cheaply. If the information is, therefore, transmitted to one buyer, he can in turn sell it very cheaply, so that the market price is well below the cost of production. But if the transmission costs are high, then it is also true that there is inappropriability, since the seller cannot realize the social value of the information. Both cases occur in practice with different kinds of information.

But then, according to well-known principles of welfare economics, the inappropriability of a commodity means that its production will be far from optimal. It may be below optimal: it may also induce costly protective measures outside the usual property system.

Thus, it has been a classic position that a competitive world will underinvest in research and development, because the information acquired will become general knowledge and cannot be appropriated by the firm financing the research. . . . if secrecy is possible, there may be overinvestment in information gathering, each firm may secretly get the same information, either on nature or on each other, although it would of course consume less of society's resources if they were collected once and disseminated to all.[13]

---

13. K. A. Arrow, *The Economics of Information* (Cambridge: Belknap Press of Harvard University Press, 1984), pp. 142–43.

If agents were omniscient, these problems would not emerge. Genomes are omniscient, so the parallel problems do not emerge in nature and do not obstruct equilibria. There are no apparent economies of scale operating within species in the reproductive fitness. And beside, the information which the environment provides about relative adaptedness is costless and universally available, so there is no problem about appropriability. In the absence of secrecy and the need for strategic knowledge about what other agents know, there is no room in biological evolution for the sort of problems information raises in economics. Once biological systems become social and their interactions become strategic, the role for information becomes crucial. But at this point evolution turns Lamarckian. It is no surprise that when "acquired" characteristics are available for differential transmission, markets for the characteristics will emerge. But at this point Darwinian evolution is no longer operating. In fact, one good argument against the adoption of Darwinian evolutionary theory as a model for economic theory is just the difference made by information. Once it appears in nature, evolution ceases to be exclusively or even mainly Darwinian. Why suppose that once information becomes as important as it is in economic exchange that phenomena should again become Darwinian?

CONCLUSION

Rational agents are biological systems. In this long chapter we have examined whether treating the theory of rational choice as a theory about biological systems holds out any hope of improving or underwriting its claims and whether it provides a useful metaphor or a literal avenue of development for economic theory. The answer does not appear to be in the affirmative. Together with economists' resolute commitment against taking psychology seriously, there seem no avenues along which to link it with the rest of science—social or natural. The absence of such natural links makes it worthwhile to consider whether economic theory should be viewed as an explanatory and predictive enterprise at all, or at least as one reconciled to at most generic prediction. The next chapter, on the role of equilibrium analysis in the theory, makes this conclusion more attractive.

7

WHY GENERAL
EQUILIBRIUM THEORY?

Along with an unshakable attachment to the formalization of common sense as an explanatory theory, neoclassical economics has formed an equally irresistible attraction to the goal of explaining phenomena by deriving their existence as instances of an equilibrium, and if possible a unique and stable one at that. The fact that few economic processes appear to cry out for such treatment does little to distract economists from this goal. In this chapter, I examine the motivation behind this attachment and whether there is anything that does or could justify it. Understanding the grip of general equilibrium on economists, together with the grip intentional variables have, should explain much about the cognitive status of economic theory.

"ECONOMICS JUST IS GENERAL EQUILIBRIUM ANALYSIS"

Given the announced attachment of economic method to falsifiability as a methodological stricture, the unmovable position of general equilibrium as a theoretical desideratum should come as something of a shock. More than any other style of theorizing, equilibrium claims are least open to being falsified, at least in theory.

The existence and stability of a general equilibrium are like the unicorn, things we could go on looking for forever without falsifying the hypotheses that there are such things. Of course, once we have established that a general equilibrium exists and that it is stable, falsifying the claim that it is unique is much less difficult: simply find another set of prices at which all markets clear and we have falsified the claim of uniqueness. The falsifiability of the claim of uniqueness is a small comfort in the circumstances.

But if the chief theoretical ornament of economic theory is in principle unfalsifiable when treated as a contingent hypothesis, to what can we attribute the attachment scientific economists feel for it? There are of course claims about ideological causes, as well as sociological, historical, or biographical explanations, not to mention rhetorical explanations for the prominence of general equilibrium analysis. What we need, however, is an epistemological or a methodological explanation that identifies a goal unattainable other than by constructing general equilibrium models.

It will not suffice to answer this question by appealing to the place of general equilibrium in the hard core of a Lakatosian research program, even if there is such a thing. Our question is prior to the question the Lakatosian claim answers: the question, Why don't economists give up general equilibrium thinking very easily? Our question is, Why *should* they find it attractive in the first place? Although we can all agree that general equilibrium analysis constitutes the hard core of what seems to satisfy Lakatos's characterization of a scientific research program, the question to be examined is why it should be chosen to do this honor. This question is not the historical question of why it was chosen. Presumably that question is answered by appeal to the intellectual biography of Walras and the influence he has exercised on twentieth-century economics. Blaug suggests that before Walras, economists

> were still capable of asking: Are prices first determined in the market by demand and supply and then passed on to consumers to permit them to reach an optimum quantity adjustment, or do consumers first decide how much to purchase and do these decisions result in market demand prices? Even if we start with given factor supplies and fixed input coefficients of production, factor prices are not determined until firms have decided to produce certain levels of output: but this decision implies knowledge of product prices, and these are not determined until households have

received income from the sales of factor services at certain prices. Obviously product and factor prices are determined simultaneously. Many contemporaries of Walras found this proposition difficult to comprehend. They never quite overcame the suspicion that the argument constituted a vicious circle. They understood the validity of partial equilibrium analysis, based on the assumption that certain variables in the analysis are treated as parameters, but they could not grasp the idea that the existence of n partial equilibria each involving n variables did not in any way guarantee general equilibrium for the whole economy made up of n markets.[1]

It is not obvious that all prices are in fact determined simultaneously, but what was needed was an account of how they possibly could be, for without such a possibility, economic theory would apparently be mired in an infinite regress. So Walras took the first steps toward showing how it was possible for all these processes to operate simultaneously and interact with one another. What he did was solve a "how possible" problem. The question remains as to why this solution to a "how possible" problem became the touchstone of theory in the discipline of economics, why it should have become an adequacy criterion on "why actual" explanations of what is really happening in a market, industry, or whole economy.

Blaug's answer to this question is that there was no reason for it to become such a criterion at all:

> [W]hat may be doubted is the notion that [general equilibrium theory] provides a fruitful starting point from which to approach a substantive explanation of the workings of an economic system. Its leading characteristic has been the endless formalization of purely logical problems without the slightest regard for the production of falsifiable theorems about actual economic behavior, which, we insist, remains the fundamental task of economics. The widespread belief that every economic theory must be fit-

1. *Economic Theory in Retrospect*, 3d ed. (Cambridge: Cambridge University Press, 1978), p. 603.

ted into the G[eneral] E[quilibrium] mold if it is to qualify as rigorous science has perhaps been more responsible than any other intellectual force for the purely abstract and nonempirical character of so much of modern economic reasoning.[2]

But if this diagnosis is right, then the mystery of why general equilibrium analysis has assumed this role is even greater.

Let's put aside the simplest and most tendentious answer: that not only does general equilibrium theory show how something is possible, it also explains the actuality of that possibility. It would be nice if we had good evidence that a general equilibrium obtains. The answer that general equilibrium obtains is the most obvious and, if it were accepted, would put to rest the question of why pursue general equilibrium analysis. If the hypothesis that there is a unique, stable, market-clearing general equilibrium were obviously well confirmed, then it would be found attractive for the best of all reasons: it would be our best guess as to what the truth is about our economy. Alas, the jury is still out on this issue, to say the least.

In fact, the most vocal defenders of the role and significance of general equilibrium thinking in economics are among the first to admit that the importance of general equilibrium is not a matter of the factual adequacy of the theory. Their justification for pursuing general equilibrium strategies in economics involves a sort of persuasive definition of economics: economics is just what we can extract from the equilibrium approach. Thus in *Equilibrium and Macroeconomics,* Hahn repeatedly says that the equilibrium notion "serves to make precise the limits of economic analysis," that when we have translated some action into "equilibrium behavior," we have gone as far as economists can in the present state of knowledge go.[3] If equilibrium thinking defines economic theorizing, general equilibrium is afforded ipso facto a central role as the limiting case of this approach and perhaps even more, as the necessary condition for partial equilibrium analysis (more on this later).

2. *The Methodology of Economics* (Cambridge: Cambridge University Press, 1980), p. 192.
3. (Oxford: Basil Blackwell, 1984), pp. 56, 67.

Defining economics as equilibrium thinking is not much more than a bald version of the Lakatosian claim that general equilibrium is the hard core of a research program. The mere fact that equilibrium thinking defines economics is no reason to suppose it is a fruitful approach to any real problem. If there were anything to this argument, a similar one could be erected to justify an obsession with demons on the grounds that without it demonology is impossible. If without equilibrium "economics" is impossible, it does not follow that the problems that "economics" must deal with disappear. And so long as they remain, there is a discipline that concerns itself with them, whether we recognize it as economics or under another name.

Another approach invokes the success of other equilibrium theories and urges economics to emulate the disciplines in which such theories have succeeded, on the grounds that what has worked so well in physics or biology should be given a good run for its money in economics. Well, there is something to this argument, especially when we tease it away from the versions associated with mere physics-envy.

The fact that a successful physical theory is cast in a certain form is no good reason in and of itself to cast an economic theory in that form. After all, there is no reason to suppose that economic phenomena are like physical phenomena, and the data of each domain ultimately should arbitrate the nature of theory in each. To let styles of theorizing in physics dictate economic styles is to let the motion of pendulums, cannonballs, and comets dictate our best guesses about the choices of consumers, the motion of markets, and the shape of economic theory.

On the other hand, as noted in the last chapter, there are reasons why physicists and biologists have found equilibrium strategies attractive, independent of the fact that they have worked predictively. In the case of biology at least the attraction must have been largely independent of the predictive success of the theory, since by itself evolutionary theory has little predictive success. These independent reasons for the attractiveness of equilibrium explanations are so general they might almost seem metaphysical.

How can we explain change? One answer to this question that has attracted attention literally since Plato's time is that change is explained when it is revealed to be merely apparent, when we can show that underneath the change, there is some unchanging fact that accounts for it. An Enlightenment version of this idea is that the persistence of arrangements over time without change is in and by itself intelligible and not in need of explanation. Accordingly, such persistence can serve to explain change, if we can show that apparent change is really the result of the persistence. As J. L. Mackie writes, in his qualified defense of this view,

> A match is struck on a matchbox and a flame appears: on the face of it this effect has nothing in common with its cause. But if we were to replace the macroscopic picture with a detailed description of the molecular and atomic movements with which the perceived processes are identified by an adequate physico-chemical theory, we should find far more continuity and persistence. . . . What is called a causal mechanism is a process which underlies a regular sequence and each phase of which exhibits qualitative as well as spatio-temporal persistence.[4]

I do not mean to endorse this view, only to note that it is a widespread, though rarely expressed, presumption that still operates in our search for explanations. General equilibrium approaches, whether in physical science, biological science, or social science, have the virtue of satisfying this presumption most fully. In a state of general equilibrium there is no change to be explained; there is just persistence. And if systems remain in equilibrium or tend toward it, then the amount of change that requires explanation is minimized.

All this makes equilibrium explanations particularly attractive independent of their success in other disciplines and independent of the prospects of harnessing them to well-worn mathematical techniques drawn from differential calculus. They provide a natural stopping point for inquiry and so an obvious target in theory construction. Their role as a stopping

4. *The Cement of the Universe* (Oxford: Oxford University Press, 1974), p. 222.

point is the strongest purely intellectual source of the attraction of equilibrium theories in the sciences, both natural and social.

These considerations, however, explain only why economists might in fact emulate other scientists in initially seeking equilibrium explanations of the ceaseless change that appears to characterize economic phenomena. At most they justify the attempt to find an equilibrium explanation for the apparent chaos of economic exchange. Nevertheless, they will not justify general equilibrium theorizing as a long-term research program. Only empirically substantiating the equilibrium explanations and improving their predictive power will do this, and that is just what is missing from general equilibrium theory. Thus, a purely methodological argument for general equilibrium theory from the nature of equilibrium explanations is insufficient.

CAN WE HAVE PARTIAL EQUILIBRIUM WITHOUT GENERAL EQUILIBRIUM?

There are two recent arguments for the cognitive importance of general equilibrium theory in economics that do not depend on any controversial prior claims about its evidential basis or methodological provenance. They seem to be about the best around, and they have the special advantage of being offered by economists whose credentials as general equilibrium theorists are unimpeachable. First, there is Frank Hahn's claim that Arrow-Debreu equilibrium has a negative role that by itself is, in Hahn's words, "almost" a "sufficient justification" for it: Arrow and Debreu showed that for a general market-clearing equilibrium, it is necessary to have futures markets in every commodity, where commodities are distinguished from one another by date and place of availability as well as by the usual differences. But no such futures market obtains, except for a few stock options and bulk commodities. And even in these instances, the markets hardly have the fineness of grain Arrow and Debreu require. After all, you can only trade in pork bellies deliverable at one place and just four times a year, and you cannot buy or sell less than 20,000 pounds. Now, according to Hahn, this is important, because pointing it out will forestall jejune claims that market prices will effectively ration nonrenewable resources.

> The argument here will turn on the absence of futures markets and contingent futures markets and on the inadequate treatment of time and uncertainty by the [Arrow-Debreu] construction. This negative role of the Arrow-Debreu equilibrium I consider almost to be sufficient justification for it, since practical men and ill-trained economists everywhere in the world do not understand what they are claiming to be the case, when they claim a beneficent and coherent role for the invisible hand.[5]

The absence of futures markets is really not much of an argument for the role general equilibrium analysis plays in modern economic theory. After all, even without it, we can explain what is wrong with the claim that price will allocate scarce goods with perfect efficiency. The absence of a complete set of contingent futures markets is in fact well down on the list of reasons why market prices are not market clearing.

A purely negative role for general equilibrium theory no more justifies its relevance to an issue like the future availability of a resource like oil than the claim that the special theory of relativity is relevant to aerodynamics on the grounds that it sets a limit on the maximum velocity of airplanes. If this were all that the special theory of relativity did, it would not be enough to sustain its centrality to physics. Mutatis mutandis for general equilibrium theory.

Hahn's justification is if anything an irony, because as Alan Nelson has noted,[6] the allocation claim refuted by careful general equilibrium analysis is almost always one advanced on the basis of the assumption that general equilibrium theory is at

5. *Equilibrium and Macroeconomics*, p. 52. It is hard to know exactly how much force to attach to this argument. On the one hand, it is not clear what it means for a justification to be almost sufficient. On the other hand, Hahn goes on to write, "But for descriptive purposes this negative role is hardly a recommendation." Which side is Hahn on?

6. "Review of Hahn, *Equilibrium and Macroeconomics*," *Economics and Philosophy* 2 (1986): 149.

least approximately right as a claim about the workings of the economy. It is hardly a justification of the theory when we use it to show the theory's boundary conditions are *not* realized by any real economy.

Another unconvincing reason for the importance of general equilibrium is offered by Weintraub.[7] Roughly, the argument is that partial equilibrium analysis is impossible except against the background of an assurance of general equilibrium. Comparative statics involves inferring from a change in one variable, like demand, to another variable, like price, or in the case of complements or substitutes, from the price of one to the price of another. But we can make no such inference without the assurance that the first change leaves other variables unaffected or affects them in ways that cancel out in compensating ways, and without the assurance that the second change will not affect other prices in an avalanche of cascading ways that ultimately undermine the entire price system, so that comparative statics is impossible.

Weintraub's argument is a double-edged sword, however. If it is right, it will also support the claim that if we know that the conditions for a market-clearing general equilibrium do not obtain, then perhaps even so innocent an enterprise as comparative statics is compromised. But if this argument is a double-edged sword, it is one with blunt edges. The fact is we have a great deal of empirical, anecdotal, factual, relatively non-theoretical evidence that the price system is stable enough that changes in the price of one commodity will not send ever-increasing effects throughout the economy leading to some catastrophic explosion of prices. Even given money illusions and hyperinflations, relative prices remain relatively constant, despite the uncertainties produced. It would of course be nice to have a theoretical assurance that partial equilibrium analysis

---

7. *General Equilibrium Analysis* (Cambridge: Cambridge University Press, 1985). Perhaps Weintraub does not fully endorse the argument to follow, for he puts it into the mouth of one of the interlocutors in a dialogue. The fact that the interlocutor who offers this argument is the economist and the others are his students emboldens me to attribute it to him.

is possible, but in its absence there is still considerable reason for confidence in its employment.

Indeed, the shoe may be on the other foot. That is, according to some accounts, partial equilibrium analysis is the more basic or fundamental component of economic theorizing, and general equilibrium theory is the derivative application. Daniel Hausman has long argued that what he calls "equilibrium theory" is the fundamental theory in microeconomics: it is composed of the recognizably basic assumptions about consumers and producers that every economics text recounts, plus assumptions about decreasing returns to scale, the convexity of production functions, and the assumption that entrepreneurs maximize profits.[8] According to Hausman the application of these assumptions to undertake comparative static analysis requires only that markets be relatively independent of one another—a fairly reasonable assumption. Such analysis constitutes a "partial equilibrium model."

Now, if we add *some further* assumptions, restrictions, or complications to the basic ones of "equilibrium theory," the result must be a derivative theory, not a more fundamental one, or so Hausman claims. And indeed, partial and general equilibrium theories are both derivative in this sense; the latter involves applying the basic components of microeconomics under two such further assumptions: that there are many commodities and that there is a general interdependence among all markets.

If Hausman is right, then not only is general equilibrium theory not required to underwrite partial equilibrium analysis, but they are both further specifications on another distinct and more fundamental theory, equilibrium theory, which we need to start with if we are to have a hope of proving general equilibrium.

I am inclined to think that this counterargument is inconclusive, however. It is by no means clear that equilibrium theory *sans phrase* makes no assumptions about the independence or interdependence of prices. The last of nine "lawlike state-

8. See *Capital, Prices and Profits* (New York: Columbia University Press), 1981.

ments" which Hausman takes to constitute equilibrium theory is "through exchange the economic choices of individuals become compatible."[9] This assumption of equilibrium theory will be true only if prices are so related as to provide for stability, and this in turn will obtain only if they are independent of one another—as partial equilibrium assumes and market clearing, as general equilibrium, presumes. If one argues that independence of prices is more basic a case than partial or complete interdependence of them, then partial equilibrium turns out to be more basic than general equilibrium, pace Weintraub. But if independence of all markets from one another is a special case from a general range of possibilities, such that we can prove existence, stability, and uniqueness of equilibria for the most extreme case of complete interdependence and for all less extreme cases of partial interdependence down to the null case, then complete interdependence will be no less basic an assumption than complete independence. Indeed, proofs for the former case will be more general and in effect include the latter case as a degenerate case. So, it is not clear either that there is some third, more basic sense of equilibrium besides partial and general or which of them is really a further restriction on the other. General equilibrium may after all be a special case of partial equilibrium, or partial equilibrium a restriction on the general case. Neither of these alternatives will vindicate Weintraub's argument, for, Hausman rightly argues, no matter whether derivative or fundamental, general equilibrium proofs have no explanatory role, because there is no equilibrium to explain. The notion that general equilibrium analysis should have a nonexplanatory role in economics is a point to which I return below.

In other disciplines, partial equilibrium analyses are conducted not only in the absence of any assurance of general equilibrium but in the open recognition that no such equilibrium obtains. For example, in evolutionary theory, we can trace the effects on the gene pool of the immigration of new genotypes by comparing gene frequencies before and after immigration. We can also be sure that for every immigration into a gene pool

9. Ibid., p. 109.

there is an equal emigration from another gene pool, with a concomitant change in its gene frequencies. It is always theoretically possible that this latter change will have further ramifications on the equilibrium level of the former population, preventing it from reaching any stable equilibrium. But the theoretical possibility of instability provides no real basis for anxiety about local genetic equilibrium. Similarly, in Newtonian mechanics, the effects of a disturbance on a system are calculated on the assumption that it is "closed." For example, position and momentum calculations for the Solar System are made on the assumption that the distant stars are fixed in their positions and unaffected by the movement of the Solar System. Yet, we know that because gravitational forces are transmitted instantaneously and infinitely this assumption is false. Nevertheless, observation is sufficient to assure us that no untoward consequences follow from the assumption that the disturbances' effects are not propagated outward in such a way as to result in chaos.

WHAT NEEDS EXPLANATION?
CAN GENERAL EQUILIBRIUM
THEORY EXPLAIN IT?

Most discussions of general equilibrium theory begin with the assumption that, like any theory which proceeds as an existence proof, the objective of general equilibrium theory is to explain why there is a unique stable market-clearing equilibrium, and do so by proving its existence. Since it is, to say the least, doubtful that such an equilibrium exists, perplexity about the cognitive status and systematic role of general equilibrium theory emerges. Thus, solutions to the problem of explaining why general equilibrium theory is central to economics presuppose two things: that, like any scientific theory, general equilibrium theory has an explanatory function, and that its explanandum phenomenon is the equilibrium whose existence it establishes.

Both of these assumptions can be questioned. Elsewhere I have challenged the second assumption suggesting that the cognitive status of general equilibrium theory might become clearer if we realized that its explanandum was not equilibrium but something else: *stability* of the price system.[10] Part of the reason to seek an alternative explanandum for general equilibrium theory is of course the fact that the existence of a general equilibrium is completely unsuitable as the explanandum phenomenon of general equilibrium theory. First of all, as is obvious to everyone except a few rational expectation theorists, there is no general equilibrium to explain, still less a market-clearing one. Second, general equilibrium theory embodies more than just the existence of a market-clearing price vector. There are the proofs that the equilibrium this price vector generates is unique and that it is a stable price vector: changes in the price of one good will generate compensating changes in the rest of the market that bring the economy back to the equilibrium level of prices. Finally, general equilibrium theory has important welfare-economic consequences, of which more later. All in all general equilibrium theory is more than an explanation for something that doesn't occur.

Now, the existence and uniqueness of equilibrium are highly theoretical matters, about which there is much debate, so much that the claim they constitute the explanandum phenomenon of general equilibrium theory must be suspect. We normally demand explanations for events or states we know actually to occur. Hypothesized facts, like the existence and uniqueness of general equilibrium, are usually to be found in the explanans, not the explanandum, of an explanation. For instance, in evolutionary biology, the existence and stability of an equilibrium figure as part of the explanation for observed adaptations and their evolutionary stability. Here equilibrium is not a fact to be explained. It is a claim that does the explaining.

But the same cannot be said of the stability of prices. Here we have an explanandum phenomenon that all parties agree obtains and that needs explanation. The fact is that though prices are interconnected, the price system is not unstable: when one

10. See my article "The Explanatory Role of Existence Proofs," *Ethics* 97 (1986): 177–87. The idea that price stability is the explanandum phenomenon of economic theory is a notion that Weintraub explores in *Stabilizing Dynamics: Constructing Economic Knowledge* (Cambridge: Cambridge University Press, 1991).

price changes, the result is not a catastrophic collapse of the whole set of interrelationships; changes do not accelerate and feed back on themselves in such a way as to destroy the basis for exchange; changes in one market do not send shock waves through all other markets, though they do produce readjustments in them. As noted above, even in hyperinflations, the relative prices among commodities remain remarkably stable. Stability demands explanation. So, perhaps we can treat the existence of price stability as the explanandum phenomenon of general equilibrium theory and view the proofs of existence and uniqueness as parts of an explanation of this fact of stability.

The overarching importance of price stability for according content to general equilibrium theory is illustrated by passages like the following from Quirk and Saposnik's *Introduction to General Equilibrium and Welfare Economics:* "Stability of equilibrium positions occupies a special place in the study of comparative statics. This is because comparative static theorems, which are statements about changes in the equilibrium values of economic variables when equilibrium is disturbed, have no predictive content if equilibrium is dynamically unstable."[11] But if stability is what gives general equilibrium predictive content and thus is something that needs to be explained, what role do existence and uniqueness of a general equilibrium play in an explanation? How can a theory which turns on their existence explain what does obtain, when they do not obtain?

One possible answer depends on the notion that a theory is explanatory if it describes a process leading to the fact to be explained that *would* produce what actually obtains—price stability—were it not *preempted* by other processes that do produce stability.[12] But can we make much of a case for viewing general equilibrium theory as part of a preempted process that

---

11. J. Quirk and R. Saposnik, *Introduction to General Equilibrium and Welfare Economics* (New York: McGraw-Hill, 1968), pp. 198–99.

12. The idea derives from Robert Nozick, *Anarchy, State and Utopia* (New York: Basic Books, 1974), and is explored in Alan Nelson, "Explanation and Justification in Political Philosophy," *Ethics* 97 (1986): 154–76. My paper "The Explanatory Role of Existence Proofs" began as a comment on Nelson and got out of hand.

begins with arbitrary endowments and ends with price stability via equilibration over time? So viewing general equilibrium theory may seem doubtful. After all, it is a persistent criticism of general equilibrium theory that it provides no coherent theory of the process by which equilibrium originates.[13] The Walrasian explanation of equilibrium in terms of *tâtonnement* cannot be viewed as a series of temporally sequential trades from initial positions through a succession of exchanges to a final, equilibrating one. Nonequilibrium trading could just as well keep altering the equilibrium solution instead of converging toward it. In Walrasian *tâtonnement* there is no process of approximation to equilibrium; nothing happens until the price vector that clears all markets is found. Only then does exchange take place. There is nothing in this that could be a preempted but possible history through which price stability originated. The situation envisioned by *tâtonnement* is one of bargaining behind a veil of ignorance until the participants choose the price vector the economist hopes for.[14]

Even if we assume, along with the exponents of general equilibrium theory, that there is at least in principle some way to realize the price vector that current theory can only identify in a *tâtonnement* process, some sort of Edgworthian groping process that finds its way toward equilibrium and keeps us in its vicinity, the most we are going to have is a preempted explanation. Why? Because this groping operates to attain stability through the vehicle of achieving a unique equilibrium. And there is no such thing. That's why we say the general equilibrium theory mechanism is preempted.

How much use can a preempted explanation be? Exponents of the centrality of general equilibrium theory are going to have

13. On this point see, for instance, M. Blaug, *Economic Theory in Retrospect,* 3d ed. (Cambridge: Cambridge University Press, 1978), pp. 610–12.

14. There are, of course, theorems in general equilibrium theory to the effect that under certain circumstances non-*tâtonnement* processes involving a sequence of nonequilibrium exchanges beginning with arbitrary endowments can converge on an equilibrium. See, for instance, F. Hahn and T. Negishi, "A Theorem on Non-*Tâtonnement* Stability," *Econometrica* 30 (1962): 463–69. But these theorems do not guarantee stability, only an equilibrium and only under restrictive conditions.

to give a very strong and implausible answer to this question. They will have to say that though preempted and bereft of a real mechanism to secure stability, this theory is the most central part of neoclassical economics. To make this claim they will have to show how in general a preempted explanation can be explanatory at all. A preempted explanation may be a useful predictive device and as such be an important scientific accomplishment, like Ptolemaic astronomy, but general equilibrium theory certainly can make no such claim. On the other hand, preempted theories are generally viewed as nonexplanatory just because the processes they envision have been preempted by other ones. After all, it is easy to dream up alternative scenarios about how we get from one state of affairs to another, but no such scenarios have explanatory power unless they bear a more robust relation to the actual scenario than the relation of being preempted by it.

What this suggests is that even changing the explanandum phenomenon of general equilibrium theory so that it is aimed at explaining something that really obtains, stability, instead of something that does not, equilibrium, does not seem able to circumvent the problems facing the theory.

So if the preempted mechanisms of general equilibrium theory are going to be of any interest to us, we need to find another function for them to serve. That is, we need to challenge the assumption that they have an explanatory purpose to begin with.

GENERAL EQUILIBRIUM AND THE SOCIAL CONTRACT

Our discussion of general equilibrium theory and its content has so far been silent on the welfare-economic dimensions of the subject. Yet we know that the two theorems of welfare economics that can be extracted from general equilibrium theory are a major intellectual achievement and a powerful reflection of its charms for economists. On the other hand, it behooves the economist to remain silent on the role of these theorems in the context of any inquiry into the explanatory role of general equilibrium theory. At most they show that if one endorses a certain distribution of commodities, one can secure it through a certain market mechanism; and if one establishes a market mechanism among rational consumers, one can be sure that the result will be desirable according to at least one very weak normative principle of desirability. The former claim is a hypothetical imperative and therefore without explanatory content. The latter is a conditional claim about the attainment of a morally approved goal ceteris paribus, and its relevance to the real world is suspect for the same reason the theory as a whole is. Neither will make much of a contribution to justifying general equilibrium theory as an explanatory enterprise, nor will other normative considerations.

However, now we seek a nonexplanatory rationale for the centrality of general equilibrium theory in economic theory. Let us consider whether the rationale might in the end be normative. This is not a particularly new idea in the philosophy of social science or even in the philosophy of economics. It has often been held that theories in the social disciplines are normative because they are about action or otherwise use vocabulary that is evaluative. For example, it is often claimed that economics is a body of prescriptions on how to be rational, and rationality is taken to be a normative concept. So viewed, of course economic theory, or any normative theory for that matter, is beyond the criticism that it is predictively weak or otherwise defective as empirical science. This is little consolation, however, because normative claims are also irrelevant to the explanation of actual behavior. Thus, there is typically little more that can be said for a normative interpretation of economic theory. We can treat its claims as hypothetical imperatives about how to be rational, but doing so doesn't really advance our understanding of the cognitive status of the theory. Furthermore, once economists make it clear that the notion of rationality with which they are concerned is purely "instrumental" and that the assessment of ends for rationality is no part of their enterprise (tastes are exogenous, De gustibus . . . , etc.), the simple-minded interpretation of the theory as normative collapses.

There is another more interesting way of arguing that economic theory is fundamentally part of a normative enterprise, one which really does shed light on its character and provides an alternative nonexplanatory goal for general equilibrium theory. In what follows I shall sketch what such an interpretation looks like. It is worth noting that the rationale it provides trades on the intellectual attractions that general equilibrium theory has held for economists, even economists initially unsympathetic to the welfare-economic hypothetical imperative derivable from the theory. And the rationale helps itself to the historically most compelling of all strategy for the establishment of political philosophies: the social contract.

One hint that this may be a useful way of approaching the aims and claims of general equilibrium theory is to observe how much political philosophy and especially contractarian political theory have been influenced by economic theory in the last two decades. Many philosophers have simply taken over the jargon and the agenda of welfare economics in order to express their problems and seek solutions to them. In particular the recognition that the establishment of cooperative institutions is a public goods problem has driven philosophers to explore economists' ways of solving such problems. Naturally enough philosophers find themselves dissatisfied with the economists' solutions—typically the unargued imposition of coercion. The dissatisfaction is in part because philosophers are asking more fundamental versions of the same question. For example, instead of asking whether market failure justifies coercion, the philosopher asks whether markets can emerge before any form of coercion is justified. This is the so-called problem of premarket market failure.[15] Perhaps philosophers' enchantment with welfare economics for their purposes reflects a belated recognition which we can also expect economists explicitly to endorse: that general equilibrium theory is a species of formal political philosophy.

What is so seductive about general equilibrium theory is that it is a formal proof of an apparently surprising possibility. As Arrow and Hahn have noted, "The immediate 'common sense' answer to the question 'What will an economy motivated by in-

15. See, for instance, Jules Coleman, "Market Contractarianism and the Unanimity Rule," *Social Philosophy and Policy* 2 (1985): 69–114.

dividual greed and controlled by a very large number of different agents look alike?' is probably: there will be chaos."[16] This response is right and it is probably the chief attraction of a centrally planned economy, one in which the planners can reconcile everyone's conflicting wants and decide on a schedule of production that meets these wants optimally, given the constraints of available resources. Let us hazard a guess that most intelligent persons thinking abstractly about the need for coordination, efficiency, and equity in a society will be attracted to some form of collectivism.[17]

Arrow and Hahn, however, note, in a tone of understatement and technicality, that the contrary is the case: "A decentralized economy motivated by self-interest and guided by price signals would be compatible with a coherent disposition of economic resources that could be regarded, in a well-defined sense, as superior to a large class of possible alternative dispositions. . . . It is important to understand how surprising this claim must be to anyone not exposed to this tradition."[18] Now, imagine a large number of rational agents who have already come to agreement about the advantages that each will accrue from the existence of a state with political authority to coerce each of them as necessary to enforce rules on which they have also agreed. I grant that we do not as yet have a good "reconstruction" of exactly what argument will convince all of our agents to concur in these rules, or exactly what rules they will concur in. That is to say, contractarian political philosophy has not yet entirely solved its own problems. So, let us assume it has solved them.

Having agreed on political rules, these individuals fall to arguing about what commercial institutions they will establish. What arrangement will recommend itself? Given the assump-

16. K. Arrow and F. Hahn, *General Competitive Analysis* (San Francisco: Holden Day, 1971), p. vii.

17. One reason to think this is right is the large number of autobiographical admissions of neoclassical economists that they were attracted to economics by the socialist vision. But as Stigler is infamous for noting, nothing makes an intelligent person an opponent of central planning faster than a good dose of microeconomics.

18. *General Competitive Analysis*, p. vi.

tion contractarian political philosophy makes, that they are rational and not altruistic in their preferences, and given some undeniable facts about information, incentives, and scarcity, it is not difficult to establish the preferability for society as a whole of decentralized market mechanisms over centralized planning ones.

Except in the case of public goods, what seems most remarkable about a market economy is that by accepting the inevitability of surpluses and shortfalls, it does better at mitigating them than a planned economy, which denies their inevitability. That is, not only does it avoid shortages and surpluses more frequently, but when they do occur, they are smaller. A market economy will be especially receptive to innovations. These it brings to market quickly. If we can prove that by arranging a scheme that will mitigate gluts and shortages to the maximum extent, we do better than we would if we simply aimed directly at eliminating them through the central collection of information and rational planning in the use of it, we shall have provided a powerful incentive for rational agents to adopt such a scheme. If we can show that there is too much information from consumers and producers for a planner to process, and too much information that consumers and producers have an incentive to hide, so that gluts and shortages will be inevitable, and that a market mechanism takes advantage of both of these facts about information, then we will have little difficulty convincing the parties to a social contract that the market is the way to go. These are the kinds of considerations that seduce intelligent young minds from socialism to capitalism. Because they work, because they actually move people, we should accord them considerable respect in any account of why general equilibrium should have any claims on our attention.

General equilibrium theory is the formalized approach to the systematic study of this claim about how the unintended consequences of uncoordinated selfishness result in the most efficient exploitation of scarce resources in the satisfaction of wants. It is of course an inquiry with many limitations, ones which so far have prevented it from securing the agreement of all students of the matter. But at least now we can understand why economists continue to lavish attention on general equilibrium theory. It is not because they think it can be improved in the direction of a descriptively and predictively accurate explanation of economic activity, but because they believe it is already part of the best contractarian argument for the adoption of the market as a social institution and, more important, stands a chance of becoming an even better one, as its assumptions are weakened, changed, and varied.

It is worth considering then whether general equilibrium theory is best viewed as one important component in the research program of contractarian political philosophy. Doing so helps explain two other features of neoclassical theory: its remarkably a priori character and the temptation felt by philosophers, among others, to view it as a body of prescriptions about rational conduct.

Among others, I have long sought an explanation of the centrality of general equilibrium theory for economics in the notion that economic theory is a branch of applied mathematics, the study of the mathematical properties of transitivity among the elements of a set. The reasons I sought such an explanation reflect the conviction that the psychological theory which serves as the natural interpretation of neoclassical assumptions about agents is a scientific dead end, as I argued in chapter 5. In spite of this fact, economists are not about to give up microeconomics. The simplest explanation of this attitude is to suppose that, like other applied mathematicians, economists are relatively unconcerned about the factual application of their most central theoretical accomplishment.

The facts that might lead one to view economic theory as a branch of applied mathematics are also accommodated by the conclusion that economic theory is an exercise in the formal development of a solution to the problem of what economic institutions will be agreed upon by agents who must enter into a contract to establish them. Any solution to the contractarian problem must be relatively a priori because it can assume little about the actual details which the parties to the contract will face after agreement is reached and lives must be carried on. Probably it can assume no more than what is assumed in neoclassical theory. And the theory will also have to disregard the dim prospects for a scientific psychology that trades on expectations and preferences, for no matter how doubtful the scientific status of these intentional concepts, individual agents calculating their own interests will certainly have to make use of them. So, the contractarian approach absorbs those very features which might suggest that neoclassical theory is best assimilated to applied mathematics.

Of course, the two interpretations of economic theory, as a component of applied mathematics and part of the research program of contractarian political philosophy, are perfectly complementary. They may jointly overdetermine the centrality of general equilibrium theory to the enterprise of economics.

Contractarianism has one advantage over applied mathematics as a justification for the fascination general equilibrium theory has exercised for economists. Whereas the appeal of a problem in applied mathematics is relatively limited, the problem of choosing the optimal institution for our economic lives is a problem that should be of interest to everyone except Robinson Crusoe. Accordingly, a theory that stands a chance of providing the firmest theoretical foundations for the claim that one particular choice is the best one *should* certainly be the focus of attention that general equilibrium theory has in fact become. For now, at any rate, I am convinced that this is the best answer we can give to the question of why general equilibrium theory should be the hard core of the research program that constitutes neoclassical economics.

In effect, this approach to general equilibrium theory assimilates neoclassical economics to what James Buchanan identifies as one of its subdivisions: "constitutional economics." Buchanan distinguishes "orthodox economic analysis" and "constitutional economic analysis": "Orthodox economic analysis, whether this be interpreted in Marshallian or Walrasian terms, attempts to explain the choices of economic agents, their interactions with one another, and the results of these interactions." So far, we must disagree with Buchanan, for interpreted as an explanatory project, orthodox economic analysis, or its core at any rate, just is not explanatory. But Buchanan goes on:

> By contrast . . . , constitutional economic analysis attempts to explain the working properties of alternative sets of

legal-institutional-political rules that constrain the choices
and activities of economic agents, the rules that define the
framework within which the ordinary choices of economic
and political agents are made.

. . . the whole exercise is aimed at offering guidance to
those who participate in discussions of constitutional
change. . . . constitutional economics offers a potential for
normative advice to members of the constitutional convention. . . . it examines the *choice of constraints* as opposed to
the *choice within constraints.*[19]

But to decide on choice of constraints, we need information
about the effects of choice within those constraints. To the extent that orthodox economics provides information about those
choices within constraints, it will be a compartment of what
Buchanan calls orthodox economic analysis. But suppose that
the *only* information it reliably provides is information relevant
to the choice within constraints, that is, to what rules rational
agents would contract, were they required to do so. Then, there
would be no other role for what Buchanan calls orthodox economic analysis.

Now, the doubts about the explanatory relevance of general
equilibrium theory suggest that it cannot explain choice within
constraints. This is, so to speak, how the problem of justifying
general equilibrium theory starts. But these doubts are irrelevant to its role in a search for optimal constraints, for in this
search there is every reason to assume that agents ruthlessly
maximize their utilities everywhere and always, that they dis-
simulate when it is to their advantage, and that they free ride
where they can. This is just the sort of behavior against which a
polity must protect itself. Therefore, as Hume writes (and
Buchanan is fond of quoting): "Political writers have established
it as a maxim that, in contriving any system of government, and
fixing several checks and controuls of the constitution, every
man ought to be supposed a *knave,* and to have no other end, in

19. Buchanan, "Constitutional Economics," in *Explorations into Constitutional Economics* (College Station: Texas A & M University Press, 1989), p. 64.

all his actions, than private interest."[20] There is a prudential requirement that for purposes of institution design we treat all agents as utility maximizers and assure ourselves that each of us is willing to live with the consequences of doing so. The prudential requirement is honored in an especially clear and powerful way by general equilibrium theory, as its game-theoretical development reveals.

As Weintraub's history makes clear,[21] it was early recognized that the existence of a general equilibrium was also the solution to an n-person generalization of the Von Neumann–Morgenstern solution for a two-person zero-sum competitive game. Arrow and Debreu's version of this realization is particularly instructive for the assimilation of general equilibrium to a contractarian agenda.[22] We are given two kinds of agents: price-taking budget-constrained utility maximizers and an omniscient auctioneer, whose only aim is to minimize excess demand among the other agents. The second agent announces price vectors, for which the other players each announce their best response. The auctioneer chooses the one that clears the markets most fully. His position is little different from that of Hobbes's sovereign. The auctioneer and his institution are thus shown to be the most advantageous arrangement that parties to a contract about the rules for economic activity can adopt in the abstract.

The equilibrium whose existence it establishes can be shown to be a Nash equilibrium—one in which each egoistical agent has an optimal strategy, regardless of the strategies of other agents. The market in which an equilibrium-producing price vector exists thus has an especially desirable property from the point of view of contractarian political philosophy: no one can do better by adopting another strategy, and the strategies result in an informationally efficient, market-clearing, Pareto-optimal, unique, stable equilibrium.

20. David Hume, "On the Independency of Parliament," in *Essays, Moral, Political, and Literary* (Liberty Classics, 1985), p. 42.

21. See *General Equilibrium Analysis*, chap. 6.

22. "Existence of an Equilibrium for a Competitive Economy," *Econometrica* 22 (1954): 376–86.

It remains possible to cavil at the abstractness of general equilibrium theory, at the demanding assumptions required to prove that there is a market-clearing price vector. But the evident desirability of such a price vector makes it worth attempting to identify the institutional constraints under which the equilibrium such a vector produces is attainable, no matter how unbridled the egoism of Hume's knave.

ARE GENERIC PREDICTIONS
ENOUGH AFTER ALL?

Can a theory that is, as I have argued, predictively weak carry the normative burden that general equilibrium is being saddled with by this interpretation? The question reflects an objection to treating general equilibrium theory as a species of political philosophy. As the solution to a set of normative problems, general equilibrium theory still needs to have some measure of relevance to actual choice. After all, "ought" implies "can" and if we ought to adopt institutions that approach those of the market that general equilibrium describes, then it must be the case that we can do so. We have no assurance of this possibility, however, unless economic theory has a certain amount of explanatory and predictive power. If the actions it counsels are beyond us, it is irrelevant as moral philosophy. If rational choices are within our abilities, then the fact that we do not seem to engage in these activities fully enough to give the theory much empirical warrant must reduce its normative bearing as well.

How much predictive power does a theory need in order to have normative bearing on institutional design? Recall the reasons Buchanan gives for adopting the prudential assumption that agents will be "knaves," in Hume's term—self-seeking but enlightened egoists. Buchanan gives us grounds to suggest that the most we need from economic theory for purposes of political philosophy is generic prediction.

As Hume noted, it is "a maxim that, in contriving any system of government, and fixing several check and controuls of the constitution," we should assume the worst-case scenario—that all agents seek only their own advantage. We should make this assumption because we cannot be sure how people will act, and we wish to avoid the worst possible outcome for ourselves, the outcome that ensues from selfishness, free riding, and non-cooperation.

If people are at least sometimes capable of such actions, and if sometimes they actually so comport themselves, then at a minimum the theory that describes the institutions we contract for, should predict this generic possibility. If it does more, then well and good. If it can give us confidence that in some specifiable cases we need not make this worst-case assumption, all the better. At a minimum the contractual arrangements justified must be compatible with the possibility of the operation of rational self-interest, and the theory from which the arrangements are derived should enable us to predict what the generic consequences of self-interest will be.

In effect, general equilibrium theory becomes a component of what economists following Hurwicz have called the search for "incentive-compatible mechanisms."[23] Given the incentive individuals have to acquire, employ, and, of course, withhold accurate information about their beliefs and preferences in order to drive better bargains, institutions intended to advantage all members of a society need to be compatible with these incentives (whence the problem of "incentive compatibility"). Adam Smith's view was that the market provided an incentive-compatible mechanism, and his followers have held that central planning fails to solve this problem of incentive compatibility. The search for incentive-compatible mechanisms of social interaction does not presume that agents always and everywhere fully respond to their incentives to misrepresent their preferences and expectations but only that they sometimes do so.

Much of the initial work on incentive compatibility involves showing that the sort of market general equilibrium theory envisions fails to meet minimal normative standards of Pareto optimality. It is well known that competitive markets fail to supply agents with the optimal level of public goods, because each individual has an incentive to misrepresent the strength of prefer-

23. "On Informationally Decentralized Systems," in *Decision and Organization,* edited by Ray Radner and Charles B. McGuire (Amsterdam: North-Holland, 1972).

ence for the good, as measured by the amount the agent states he or she would pay for a certain quantity of the public good. Similarly, when the number of agents is finite and/or others of the stringent assumptions of the proofs of general equilibrium are waived, markets are provably nonoptimal. However, all we need for these results to be of interest is that individuals at least sometimes so act as to preclude the attainment of Pareto optima. It would be wrong to infer from the provably nonoptimal character of markets with finite numbers of agents or with public goods that general equilibrium theory has been shown to be without normative significance. To begin with, the nonoptimality proofs owe their origin to attempts to increase the realism of the assumptions of general equilibrium theory; second, in cases where we deal with a very large number of agents and a very small number of unimportant public goods, the market may come close to satisfying optimality demands and so presents itself as an attractive solution to a contractarian bargaining situation.

However, Richard Lipsey and Kevin Lancaster have shown that in general when some of the conditions for a Pareto optimum are not attained, the attainment by an economy of the other conditions does not improve the degree of attainment of such an optimum.[24] In other words, to the degree that the market envisioned by general equilibrium theory is an idealization beyond complete attainment, rational contractors, knowing this fact, should not consent to institutions that will realize only some and not all of the theory's necessary conditions. This result of course limits the attractiveness of any argument for competitive markets as the outcome of a bargain, but again it requires of economic theory no more than generic prediction. It rests on derivations of whether conditions of a certain sort can or cannot obtain and not the prediction of when and to what extent they will obtain.

Under the conditions in which a Pareto optimum does not obtain, the problem remains of what institutions should be

24. "The General Theory of the Second Best," *Review of Economic Studies* 24 (1956): 11–32.

adopted. This issue has become known as the problem of the "second best."[25] If the best outcome is not attainable, how can we get as close to it as possible? Attempts to answer questions about the second best call for more than merely generic prediction; they call for detailed information about individual agents and aggregates of them—how they will respond to institutional constraints. Without such information, it is impossible to tell whether we are coming closer to or getting farther from the unattainable optimum a general equilibrium promises.

In this respect, the account of the nature and centrality of general equilibrium theory still leaves an important aim of an economic science unaccomplished. Having assigned general equilibrium theory to contractarian political philosophy, we may still need a theory of economic behavior that transcends generic prediction. Finding second bests may require a general theory that explains actual economic phenomena. The search for such a theory will be perfectly compatible with economists' continued intellectual fascination with general equilibrium theory, even if it does not prove to be the foundation of an empirical science at all.

25. James E. Meade, *Trade and Welfare* (New York: Oxford University Press, 1955).

8

IS ECONOMIC THEORY
MATHEMATICS?

In chapter 7 I suggested that there is some reason to treat economic theory as a part of political philosophy—in particular, as an area of debate in social contract theory. This is a view which some economists might be eager to adopt, especially those preoccupied by welfare economics. But others are, I suspect, inclined to find it too limited an area for the influence of their discipline. Most economists will insist that their discipline is a science, and their theories the best and most powerful explanatory and predictive devices there are among the social sciences. Given the unsettled character of the philosophy of science, it is hard to know exactly what a science is, so the first of the economists' claims, that their discipline is one, is hard to assess. As for the second, we may regretfully acquiesce to its truth, but this is more a reflection of the theoretical impoverishment of the other social sciences. However, anyone with much knowledge of the history of economic theory will agree that the discipline does not seek or respond to empirical data in the way characteristic of an empirical science—even a theoretically impoverished one. If social contract theory is too Procrustean a bed for economic theory, and empirical science too demanding a status, is there not some other interpretation of the aims and methods of economic theory that will do full justice to its scope and its insulation from data? More than once, students of the subject have sought an interpretation of its aims and claims that adequately reflects its mathematical character. Both detractors and defenders of orthodox neoclassical theory have found its intellectual core in its mathematical expression.[1] In this final chapter I want to explore this interpretation of economic theory as a branch of applied mathematics.

In exploring this interpretation it is important to bear in mind that interpreting economic theory as a compartment of mathematics aims at understanding the subject, not condemning it as cognitively defective. After all, there is no more cognitively respectable a discipline than mathematics. On the other hand, we do not make demands of mathematics beyond its well-understood capacities. Unlike defenders or detractors of economic theory, my aim is to determine the capacities of economic theory.

EXTREMAL THEORIES AND INTENTIONALITY

As noted in chapters 3 and 4, for a long time after 1945, it might confidently have been said that Keynesian macroeconomics would ultimately provide the sort of explanatory and predictive satisfaction characteristic of science. But the simultaneous inflation and unemployment of the seventies and the high interest rates, budget balance, and reduced inflation with higher employment of the eighties and the economy's apparent imperviousness to fiscal policy eroded the confidence of economists and noneconomists alike in the theory. Moreover, some of the professional economists' reaction to the failures of Keynesian theory was even more disquieting to those who viewed economic theory as unimpeachably an empirically controlled enterprise. Rational expectations theory, the most fashionable response to the alleged failure of Keynesianism, returns to the classical economic theories which Keynesianism claimed to have superseded. The diagnosis offered by rational expectations theorists for the failure of the Keynesian theory has been that it

1. Detractors include of course Wassily Leontief. For a recent example of a defender, see Gerard Debreu, "The Mathematization of Economics," *American Economic Review* 81 (1991): 1–7.

does not accord individual agents the kind of rationality in the use of information and the satisfaction of preferences that neo-classical economic theory accords agents. The theoretical alternative offered to replace Keynesianism in the light of this diagnosis is nothing more or less than a return to the status quo ante, to the classical theory of Walras, Marshall, and the early Hicks, which Keynesianism had preempted. This cycle brings economic theory right back to where it was before 1937 and might well undermine confidence that economics is an empirical science.

Of course, a century is not a long time in the life of a science, or even a theory, so the fact that economics has not substantially changed, either in its form or in its degree of confirmation since Walras or arguably since Adam Smith, is not reason to deny it scientific respectability. But it is worth asking why economics has not moved away from the theoretical strategies that have characterized it at least since 1874, in spite of their practical inapplicability to crucial matters like the business cycle, economic development, or stagflation.

In the Lakatosian view of proper scientific method, of course, economists have been doing just what they should be doing. Since the nineteenth century they have been pursuing a single research strategy, acting in accordance with the hard core of a ubiquitous and powerful paradigm. For, as I noted in chapter 5, in pursuing equilibrium results, economists have been steadily elaborating a theory whose *form* is identical to that of the great theoretical breakthroughs in science since the sixteenth century. Accordingly it may be argued that it would be irrational for economists to surrender this strategy short of a conclusive demonstration that it is inappropriate to the explanation of economic activity. The strategy in question is reflected in the joint commitment of economics to the preference/expectation model and to equilibrium thinking: between them they enable the economist to adopt the same approach to his explananda as that employed in Newtonian mechanics or evolutionary biology, arguably the two most successful approaches of post-seventeenth-century science.

It is no surprise that a strategy which serves so well in these two signal accomplishments of science should have as strong a grip in other domains to which it seems applicable. Moreover, the constraints on theoretical and empirical developments that this strategy imposes can explain many of the greatest successes of Newtonian and Darwinian science and much of the puzzling character of developments in economic theory.

I call this strategy the extremal strategy, because it is especially apparent in Newtonian mechanics when that theory is expressed in so-called extremal principles, according to which a system's behavior always minimizes or maximizes variables reflecting the mechanically possible states of the system. In the theory of natural selection this strategy assumes that the environment acts so as to maximize fitness. The extremal strategy is crucial to the success of these theories because of the way it directs and shapes the research motivated by them. Thus, if we believe that a system always acts to maximize the value of a mechanical variable, for example, total energy, and our measurements of the observable value of that variable diverge from the predictions of the theory and the initial conditions, we do not infer that the system described is failing to maximize the value of the variable in question. We do not falsify the theory. We assume that we have incompletely described the constraints under which the system is actually operating. In Newtonian mechanics attempts to more completely describe the systems under study resulted in the discovery of new planets (Neptune, Uranus, and Pluto), the invention of new instruments, and eventually the discovery of new laws, like those of thermodynamics.

Similarly in biology, assuming that fitness is maximized led to the discovery of forces not previously recognized to effect genetic variation within a population and, more important, led to the discovery of genetic laws that explain the persistence in a population of apparently maladaptive traits, like sickle-cell anemia, for instance. Because these theories are "extremal" ones, differential calculus may be employed to express and interrelate their leading ideas.

Microeconomics is an avowedly extremal theory. Like these other theories, it asserts that the systems it describes maximize a quantity—in this case, utility (or preference, etc.) instead of fitness or total energy. That is why it can be couched in the language of differential calculus. It is the extremal character of the theory and not the fact that it deals with "quantifiable" variables, like money, that made microeconomics susceptible of quantification while other social sciences struggled with equations of doubtful meaning.

More important than the fact that they all employ differential calculus (and now avail themselves of its mathematical successors in topology, for instance), theories in physics, biology, and economics are all committed to explain everything in their domains because of their extremal character. In virtue of the claim that systems in their domains always behave in a way which maximizes or minimizes some quantity, the theories ipso facto provide the explanation of all of their subjects' behavior by citing the determinants of all their subjects' relevant states. An extremal theory cannot be treated as only a partial account of the behavior of objects in its domain, or as enumerating just *some* of the many determinants of its subjects' states; for any behavior that actually fails to maximize or minimize the value of the privileged variable simply refutes the theory *tout court*.

In fact, the pervasive character of extremal theories insulates them from falsification to a degree absent from nonextremal theories. All theories are strictly unfalsifiable, simply because testing them involves auxiliary hypotheses. However, extremal theories are not only insulated against strict falsification, they are also insulated against the sort of actual falsification that usually overthrows theories, instead of auxiliary hypotheses. In the case of a nonextremal theory, falsification may lead us either to revise the auxiliary assumptions about test conditions or to revise the theory by adding new antecedent clauses to its generalizations. With extremal theories the choice is always between rejecting the auxiliary hypotheses—the description of test conditions—or rejecting the theory altogether. The only change that can be made to the theory is to deny that its subjects invariably maximize or minimize its chosen variable.

Thus high-level extremal theories like Newtonian mechanics, Darwinian evolutionary theory, and neoclassical economic theory are left untouched by apparent counterinstances. They are not simply improved by qualifications and caveats in their antecedent conditions. Apparent falsifications, certainly in the case of Newtonian and Darwinian theory, have led to new discoveries about initial conditions, like the existence of Pluto and Neptune or the surprisingly adaptive properties of the sickle-cell trait. Extremal theories are an important methodological strategy because they are so well insulated from falsification. Their insulation has enabled them to function at the core of research programs, turning what otherwise might be anomalies and counterinstances into new predictions and new opportunities for extending their domains and deepening their precision.

Accordingly, it may be argued, economists' attachment to their extremal theory represents not complacency but a well-grounded methodological conservatism. Given the fantastic successes of this approach in such diverse areas as mechanics and biology, it would be unreasonable to forgo similar strategies in the attempt to explain human behavior. So viewed, the history of attempts to make recalcitrant facts about human behavior and the economic systems humans have constructed fit the extremal theory of microeconomics reflects a commitment that is on a par with astronomers' attempts to make recalcitrant facts about planetary perihelion fit the demands of Newtonian mechanics; it is on a par with biologists' attempts to make the persistent genetic predisposition to malaria fit the facts of adaptation demanded by the theory of natural selection. Since these attempts do not discredit their theories as empty or unfalsifiable, it should not be inferred that there is anything improper in the economists' attempts to do the same thing. Or so it may be argued.

It is certainly correct that much of the commitment to microeconomic strategies does in fact reflect this sort of reasoning. After all, it is not just the intellectual prestige associated with the scope for calculus, topology, and differential geometry attending any extremal theory that explains the reluctance of economists to forgo the strategy. But the conservative rationale for the attachment of economists to extremal theories is vitiated by a crucial disanalogy between microeconomics and mechanics or evolution. Economists would indeed be well advised not to surrender their extremal research program if only they could

boast even a small part of the startling successes that other extremal research programs have achieved. But two hundred
years of work in the same direction have produced nothing
comparable to the physicists' discovery of new planets or of new
technologies by which to control the mechanical phenomena
Newton's laws systematized. Economists have attained no independently substantiated insight into their domain to rival the biologists' understanding of macroevolution and its underlying
mechanism of adaptation and heredity. There has been no signal success of economic theory akin to these advances of extremal theory. The want of such success is a disanalogy important
enough to bear explaining. Failing a satisfactory explanation,
the difference is significant enough to make us question the
economists' own received interpretation of their extremal approach and to make us query the credentials of economic theory as straightforward empirical science.

The reasons for the difference between economics and other
extremal disciplines obviously have something to do with the
fact that unlike workers in these other disciplines economists
have been unable to independently test and improve its auxiliary assumptions and have been prevented by the nature of
these assumptions from strengthening our understanding of
the theory's explanatory variables. These are two areas in which
both mechanics and evolutionary biology have made their
greatest advances.

If the character attributed by chapter 3 to the explanatory
variables of microeconomics is right, then the failure of economics to improve its predictive power and explanatory precision is not methodological or conceptual but very broadly
empirical. Despite its conceptual integrity, as an empirical theory microeconomics rests on a false but central conviction that
vitiates its assumptions and so bedevils the theorems deduced
from them. Economic theory assumes that the categories of
preference and expectation are the classes in which economic
causes are to be systematized and that the events to be explained are properly classified as actions like buying, selling,
and the movements of markets, industries, and economies that
these actions aggregate to. The theory has made this assumption because of course it is an assumption we all make about hu-

man behavior; our behavior constitutes action and is caused by the joint operation of our desires and beliefs. As we noted in chapter 5, marginalists of the late nineteenth century like Wicksteed saw clearly that microeconomics is but the formalization of this commonsense notion, and the history of the theory of consumer behavior is the search for laws that will express the relations between desire, belief, and action, first in terms of cardinal utility and certainty, later in terms of ordinal utility, revealed preference, and expected utility under varying conditions of uncertainty and risk.[2] The failure to find such a law or any approximation to it that actually improves our ability to predict consumer behavior any better than Adam Smith could have resulted on the one hand in a reinterpretation of the aims of economic theory away from explaining individual human action and on the other in the tissue of apologetics with which the consumer of economic methodology is familiar.[3]

As we have seen, the real source of trouble for the attempt to find *improvable* laws of economic behavior is something that has only become clear in the philosophy of psychology's attempts to understand the intentional variables of common sense and cognitive psychology. What we saw in chapter 5 is that "beliefs" and "desires"—the terms with which ordinary thought and the social sciences describe the causes and effects of human action—do not describe "natural kinds." They do not divide nature at the joints. They do not label types of discrete states that share the same manageably small set of causes and effects and so cannot be brought together in causal generalizations that improve on our ordinary level of prediction and control of human actions, let alone attain the sort of continuing improvement characteristic of science. We cannot expect to improve our intentional explanations of action beyond their present levels of predictive power. But the level of predictive power of our intentional theory is no higher than Plato's. The predictive weakness

2. See Alexander Rosenberg, "A Skeptical History of Economic Theory," *Theory and Decision* 12 (1980): 75–83, for a detailed account of this transformation of utility theory.

3. See Alexander Rosenberg, "Obstacles to Nomological Connection of Reasons and Actions," *Philosophy of Science* 10 (1980): 79–91.

of theories couched in intentional vocabulary reflects the fact that the terms of this vocabulary do not correlate in a manageable way with the vocabulary of other successful scientific theories; they do not divide nature at the joints, insofar as its joints are revealed in already successful theories like those of neuroscience.

The failure of microeconomic theory to uncover laws of human behavior is due to its wrongly assuming that these laws will trade in desires, beliefs, or their cognates. And the system of propositions about markets and economies that economists have constructed on the basis of its assumptions about human behavior is deprived of improving explanatory and predictive power because its assumptions cannot be improved in a way that transmits improved precision to their consequences. Thus the failure of economics as an empirical discipline is traced, not to a conceptual mistake or to the inappropriateness of extremal theories and their elegant mathematical apparatus to human action, but to a false assumption economists share with all other social scientists, indeed with everyone who has ever explained their own or others' behavior by appeal to the operation of desires and beliefs.

Just as economists have been undeterred by previous analyses of the empirical weakness of their discipline, they are unlikely to be swayed by this diagnosis either. Indeed, the persistence of economists in pursuing the extremal and intentional approach that has been conventional for well over a century suggests that nothing could make them give it up. At any rate nothing that would make empirical scientists give up a theory will make economists give up their theoretical strategy. The unwillingness to surrender this conviction leads to the conclusion that economics is not empirical science at all. Despite its appearances and the interest of some economists in applying their formalism to practical matters, this formalism does not really have the aims, nor does it make the claims, of an unequivocally empirical theory.

ECONOMICS AS APPLIED MATHEMATICS

If economics is not an empirical science, what is it? This is not just a question of nomenclature. It is a question that must be answered if we are to assess the discipline's significance by any standard other than those *internal* to economics itself. Of course many, like McCloskey, will hold that there are no such transdisciplinary standards, and others will be indifferent to how well economics measures against any standard. They are simply interested in cultivating their subject for its intrinsic appeal. The rest of us, who are neither economists nor Sophists, have no such luxury.

In chapter 7 I suggested some reasons for thinking that economic theory is a branch of political philosophy. This will explain at least a good portion of its insulation from and indifference to empirical data. But so will the suggestion that microeconomic theory is after all best understood as a division of applied mathematics. The claim that microeconomic theory, or at least its core, is a branch of abstract mathematics reflects both its differences from empirical science and its derivational, deductive structure as fully or more fully perhaps than treating it as a portion of political philosophy.

It is convenient to begin this exploration of the claim that economics is a branch of applied mathematics by considering an obvious objection to it. To begin with, it will be said, surely the fact that the fundamental axioms of economic theory fail to divide nature at the joints does not vitiate the entire enterprise. After all, this interpretation would have us forgo not just the abstract claims about preference orders and individual choice under certainty, but also the laws of supply and demand. Yet plainly these are useful approximations, regularities roughly and frequently enough instanced to reflect some underlying truth about economic behavior, enough at any rate to make economics a worthwhile pursuit even if all my claims are correct. Though I may have provided an abstract explanation of why we cannot improve the current theory indefinitely, I have not given enough reason to deprive it of its usefulness or deny it scientific standing.

The strength of the rejoinder rests on two undeniable considerations. One is that for all its infirmities, economic theory does at least sometimes *seem* to be insightful. Occasionally, qualitative predictions are borne out, and even more frequently, retrospective economic explanations of events that were unexpected, like a 15 percent reduction in the consumption of gasoline, can be given. The second consideration is more abstract but quite telling as a point in the philosophy of science. There are several scientific theories which to varying degrees fail to divide nature at the joints, and yet they are useful approximations, even if they cannot be reduced to more fundamental theories that reflect real natural kinds. For instance, the Mendelian unit of inheritance cannot be reduced to the molecular gene and so does not divide its phenomena at the joints. Yet Mendel's laws are useful approximations that we would be silly to forgo. Mutatis mutandis for economic theory. Thus, even if my claims about the intentional vocabulary of economics are correct, it will be argued that they are not sufficient reason to surrender the theory.

It is quite correct that the problems I have noted for economics have parallels in other successful scientific theories. These theories have proved successful according to standards of *improving* technological and predictive success that economics has not met. The fact that some theories that do not reveal natural kinds are to a degree incommensurable with their successors or with more fundamental theories and yet still improve our predictive success is a problem in the philosophy of science. But it is a problem different from the one facing economics. Not only have these other theories, like Mendelian genetics, been able to provide clear gains in prediction and technological control, but there has been no conceptual obstacle to uncovering the degree to which they fail to carve nature at the joints, the exact point at which they go wrong. Thus in the case of Mendelian genetics, the limits of predictive success were reached relatively early, and the explanation was sought in the underlying structure and mechanism of the gene. When the physical identification of the genetic material was finally attained, it became clear that different parts of it were responsible for mutation, for recombination, and for the production of phenotypic properties. The identification of the genetic material revealed the false assumption of Mendel's laws, that these three functions were carried out by the same kind of thing, the gene, and explained why the theory's predictions were wrong beyond a certain point. These discoveries and those of the molecular structure and chemical regulation of the genetic material enable us not only to improve on Mendel's laws, by substituting a more accurate theory for his, but also to explain why in some cases it remains perfectly acceptable to continue to employ his original theory, false assumptions and all.

What chapter 5 has shown us is that we cannot expect this denouement in the case of rational choice theory. There will be no physical localization of preferences and expectations that will enable us to trace the limits of the predictive powers of neoclassical economics, still less correct them. And the reason is that the explanatory variables of economic theory, unlike those of, say, Mendelian genetics, are not linked to physical mechanisms in a way that will enable us to discover where and how they go wrong. The relationship between beliefs, desires, and behavior just does not permit us to isolate one of these variables from the others and to sharpen up our measurements of it in ways that will lead to identification of the point at which the whole story goes wrong. So, it is not just that compared with other theories, rational choice theory cuts nature farther from the bone. It is rather that we have no idea how far from the bone it cuts, and no way to find out.

MICROECONOMICS AND EUCLIDEAN GEOMETRY

What of the successes of economic theory? How can we square my arcane philosopher's argument with the evident applicability of such staples as the laws of supply and demand? After all, it is a fact about markets in all commodities that *eventually* price will influence demand and supply in the directions that microeconomic theories of economic action dictate. Surely the influence of price on demand is an economic regularity and surely it is a consequence of individual choices, preferences, and beliefs. What will our view of microeconomic theory as mathematics make of the laws of supply and demand?

The first thing to note about such rhetorical questions is this:

although the laws of supply and demand and other market-level general statements are deduced from claims about the intentional determinants of individual actions, they are logically separable from such claims, and more important, they can be shown to follow from assumptions which are the direct denial of these general claims about rational action. From the assumption that individuals behave in purely habitual ways, always purchasing the same or the most nearly similar bundle of commodities available, no matter what the price, the law of downward-sloping demand follows, as it also does from the assumption that their purchases are all impulsively random.[4] The same can be shown for the choices of entrepreneurs. So merely rejecting the empirical claims of the extremal intentional approach to human behavior does not logically or even theoretically oblige us to surrender these "laws." On the other hand, it does help explain why we cannot sharpen the applicability of laws about economic aggregates beyond the most qualitative or generic levels or quantify the values of their parameters like elasticity or improve our foresight or hindsight in the employment of these principles.

More important, the fact that we can usefully employ false or vacuous general statements, up to certain limits, should be no mystery at all. Mathematics provides clear examples of the utility of both false and vacuous claims. The clearest instance of such restrictedly useful though false or vacuous general statements is Euclidean geometry. For millennia this axiomatic system was viewed as the science of space, and the great mystery that surrounded it was how we can have the apparently a priori knowledge of the nature of the world that the science of space, Euclidean geometry, gave us. Attempts to solve this mystery drove the creative energy of seventeenth- and eighteenth-century epistemology, especially the work of Leibniz and Kant, both of whom attempted to reconcile the necessity of geometry with its applicability in a world of apparently contingent spatial properties. The result was, with Kant, a theory of knowledge

4. Gary Becker, "Irrational Behavior and Economic Theory," *Journal of Political Economy* 70 (1962): 1–13.

that made geometry and physics synthetic truths about the world known a priori.

Since Poincaré and Einstein this problem has been largely resolved. Prior to the twentieth century, Euclidean geometry was *equivocally* interpreted. It was alternatively viewed as (*a*) a pure axiomatic system about abstract objects, one that constituted the implicit definitions of terms and was therefore a priori true, and (*b*) a body of claims about actual spatial relations among real objects in the world. The equivocation between these two interpretations in part caused Kant's problem of how synthetic propositions could be known a priori. Once distinguished, the general theory of relativity revealed that, interpreted as a theory of actual spatial relations, Euclidean geometry is false. Of course this discovery left untouched Euclidean geometry interpreted as a body of a priori truths implicitly defining the terms that figure in it.

So interpreted, pure geometry was left untouched by scientific developments, except of course to the extent that it was shown to be useless and inapplicable as a body of conventions beyond certain values of distance and mass in space. In retrospect, we can explain why no one ever noticed these facts about geometry and why before the acceptance of the general theory of relativity, Euclidean geometry was entirely satisfactory for settling empirical questions of geography, surveying, engineering, mechanics, and astronomy. The reason, of course, is that for these questions we neither needed nor had the means to make measurements fine enough to reveal the inadequacies of Euclidean geometry. When we need to improve our measurements beyond this level of fineness, in contemporary cosmology, for instance, we must forgo Euclidean geometry in favor of one or another of its non-Euclidean alternatives.

One way to describe the twentieth-century fate of Euclidean geometry is to say that its kind terms proved not to name *natural kinds:* nature diverges from the predictions of an applied Euclidean geometry, because it does not contain examples, realizations, and instances of the kind terms of that theory. There are no Euclidean triangles, as we came to learn only with the advent of another theory, the general theory of relativity, which not only revealed his fact but also explained the degree of success Euclidean geometry does in fact attain when applied to small regions of space.

Of course, economic theory has attained nothing like the success of Euclidean geometry, but the apparent applicability of some of its claims is to be explained by appeal to the same factors that explain why we can employ, for example, the Pythagorean theorem, even though there are no Euclidean triangles and no Euclidean straight lines. We can employ the laws of supply and demand, even though human beings are not economically rational agents; that is, we can employ these "laws" even though individuals do not make choices reflecting any empirical regularity governing their expectations and their intentions. We can employ them all right, but the laws of supply and demand cannot be applied with the usefulness and exactitude of the Pythagorean theorem, just because the kind terms of economic theory are different from the real kinds in which human behavior is correctly classified. And this difference is comparatively much greater than the difference between the kind terms of applied geometry and those of physics. There are no Euclidean triangles, but we know why, and we can calculate the amount of the divergence between any physical triangle and the Euclidean claims about it, because we have a physical theory to make these corrections, the very one that showed Euclidean geometry to be factually false. We can make no such improvements in the application of the laws of supply and demand; we can never do any better than apply them retrospectively or generically; we cannot specify their parameters or their exceptions, because the axiomatic system in which they figure diverges from the facts very greatly and because we have no associated theory that enables us to measure this divergence and make appropriate corrections. The lack of such a theory is a difference in kind between Euclidean geometry and economic theory. They differ in applicability only by degree; the predicates of neither pick out natural kinds; but they differ in kind because for Euclidean geometry there is a theory, physics, that enables us to correct and improve the applicability of its implications. There is no such theory that enables us to improve on the applicability of economic theory.

As we saw at the end of chapter 3, a theory, say, a version of cognitive psychology, that provides bridges from economic variables like preference and expectation to independently identifiable psychological states is, of course, logically possible. Such a theory might enable us to actually predict individual economic choices and to correct our microeconomic predictions of them when these predictions go wrong. Either it would enable us to improve microeconomics beyond the level at which it has been stuck for a hundred years, or it would show that the determinants of human behavior are so orthogonal to the theory's assumptions about them that microeconomics is best given up altogether. The fact is that no such theory is in the offing or on the horizon.

What is more, even if such a theory were available, it is not likely to actually deflect practicing economists from their intentional extremal research program. The reason is that economists and humanity in general are far more committed to viewing one another as agents, as people whose behavior is caused by desires and beliefs, than we are committed to viewing light rays as Euclidean straight lines. Before the general theory of relativity and even after, it has been hard for nonphysicists to surrender Euclid. It is even harder to surrender folk psychology. Indeed, the very prospect sounds incoherent. To surrender folk psychology is to surrender the view that there are desires and beliefs and they cause actions, but surely surrender is an action caused by desires and beliefs. The pragmatic contradiction of believing that there are no beliefs, however, is a problem for philosophy, not economic methodology.

If economists have not in fact been elaborating a contingent, empirical theory that successively improves our explanatory and predictive understanding of economic behavior, what has this notable intellectual achievement been aimed at? The parallel I have drawn with Euclidean geometry can help answer this question.

Euclidean geometry was once styled the science of space, but calling it a science did not make it one, and we have come to view advances in the axiomatization and extension of geometry as events not in science but in mathematics. Economics is often defined as the science of the distribution of scarce resources, but calling it a science does not make it one. For much of their histories, since 1800, advances in both these disciplines have consisted in improvements of deductive rigor, economy, and elegance of expression, in better axiomatizations, and in the proofs of more and more general results, without much concern for the usefulness of these results. In geometry, the fifth axiom, the postulate of the parallels, came increasingly to be the focus of attention, not because it was in doubt, but because it seemed so much more ampliative than the others.

The crisis of nineteenth-century geometry was provoked by the discovery that denying the postulate of the parallels did not generate a logically inconsistent axiomatic system. Thus the question of the cognitive status of geometry became acute. Some, following Plato, held it to be an intuitively certain body of abstract truths. Some, following Mill, held it to be a body of empirical generalizations. Others, following Kant, viewed it as a body of synthetic a priori truths. Matters were settled by distinguishing between geometry as a pure axiomatic system, composed of analytic truths about abstract objects with or without real physical instances, and geometry as an applied theory about the path of light rays, which was shown to be false for reasons given in the general theory of relativity. Moreover, the abstract and apparently pointless exercises of nineteenth-century geometers in developing non-Euclidean geometries turned out to have an altogether unexpected and important empirical role to play in helping us understand the structure of space after all: they apparently describe the real structure of space in the large. Of course, pure geometry, both Euclidean and non-Euclidean, has continued to be a subject of sustained mathematical interest, and both have had applications undreamed of eighty years ago.

Compare the history of economic theory during the same time. Unlike physical theory, or for that matter the other social sciences, economics has been subject to exactly the same conceptual pigeonholing as geometry. Some have viewed it, with Lionel Robbins, as a Platonic body of intuitively obvious, idealized, but nonetheless correct descriptions of human behavior. Others, following Ludwig von Mises, have insisted it is a Kantian body of synthetic a priori truths about rationality. Others, like the geometrical conventionalists and following T. W. Hutchison, have derided it as a body of tautologies, as a pure system of implicit definitions without any grip on the real world. Still others, following Mill, have held it to be a body of idealizations of rough empirical regularities. Finally, some, following Friedman, have treated it as an uninterpreted calculus in the way positivists treated geometry.[5]

But most economists, like most geometers, have gone about their business proving theorems and deriving results without giving much thought at all to the question of economic theory's cognitive status. For them the really important question, the one which parallels the geometer's concern about the postulate of the parallels, was whether Walras's theorem that a general market-clearing equilibrium exists, that it is stable and unique, follows from the axioms of microeconomic theory. Walras offered this result in 1874, as a formalization of Adam Smith's conviction about decentralized economies, but he was unable to give more than intuitive arguments for the theorem. It was only in 1934 that Abraham Wald provided an arduous and intricate satisfactory proof, and much work since his time has been devoted to producing more elegant, more intuitive, and more powerful proofs of new wrinkles on the theorem.[6]

Just as geometers in the nineteenth century explored the ramifications of varying the strongest assumptions of Euclidean geometry, economists have devoted great energies to varying equally crucial assumptions about the number of agents, their expectations, returns to scale, and divisibilities, and determining whether a consistent economy—a market-clearing equilibrium—will still result, will be stable, and will be unique. Their interests in this formal result are quite independent of,

5. Lionel Robbins, *An Essay on the Nature and Significance of Economic Science* (London: St. Martins, 1932); L. von Mises, *Human Action* (New Haven: Yale University Press, 1949); T. W. Hutchison, *The Significance and Basic Postulates of Economics* (London: Macmillan, 1938); and Milton Friedman, *Essays in Positive Economics* (Chicago: University of Chicago Press, 1953).

6. Leon Walras, *Elements of Pure Economics*, translated by W. Jaffe (Homewood, Ill.: Irwin, 1954); A. Wald, "On Some Systems of Equations for Mathematical Economics," *Econometrica* 19 (1951): 368–403. For a contemporary version of the proof, see Gerard Debreu, *Theory of Value* (New York: Wiley, 1959).

indeed are in spite of, the fact that its assumptions about production, distribution, and information are manifestly false. The proof of general equilibrium is the crowning achievement of mathematical economics. But just as geometry as a science faced a crisis in 1919 observations that confirmed the general theory of relativity, so too economic theory faced a crisis in the evident fact of the Great Depression. For a long time after 1929 economists lost the *conviction* that the Walrasian general equilibrium was at least a state toward which markets must, in the long run, move.

The main reaction to this crisis was, of course, Keynesianism. Insofar as this extremal theory rests on a denial of the fundamental microeconomic assumptions that economic agents' expectations are rational, that they do not suffer from money illusions, that they will tailor their actions to current and future economic environments, Keynesian theory represents as much of a conceptual revolution as non-Euclidean geometry did. Keynes did not entirely win the field, even during the period when his theory appeared to explain why the market-clearing general equilibrium might never be approached, let alone realized. One reason for this is that many economists continued to be interested in the purely formal questions of the conventional theory, quite regardless of its irrelevance to understanding the actual world. These economists were implicitly treating microeconomics as a pure axiomatic system, whose terms may or may not be instantiated in the real world, but which is of great interest, like Euclidean geometry, whether or not its objects actually exist.

More crucially for the history of economics, there never was and is not yet a theory that can play a role for economics like the role played for geometry by physical theory. Physics enables us to choose between alternative applied geometries and to explain the deviations from actual observation of the ones we reject. There is no such theory to serve as an auxiliary in any choice between an applied neoclassical equilibrium theory and a Keynesian equilibrium theory. When, in the 1970s, Keynesian theory foundered on empirical facts of joint unemployment and inflation, as unremitting as was the fact of the apparent non-market-clearing equilibrium of the 1930s, the result was an eager return to the traditional theory. Economists have not forgotten the Great Depression, but their interest in it seems limited to showing that, after all, the Walrasian approach is at least logically consistent with it, something Keynes's earliest opponents would have vouchsafed them. In short, the theory and its development have been as insulated from empirical influences as geometry ever was before Einstein. All this suggests that, like geometry, economics is best viewed as a branch of mathematics somewhere on the intersection between pure and applied axiomatic systems.

Much of the mystery surrounding the actual development of economic theory—its shifts in formalism, its insulation from empirical assessment, its interest in proving purely formal, abstract possibilities, its unchanged character over a period of centuries, the controversies about its cognitive status—can be comprehended and properly appreciated if we give up on the notion that economics any longer has the aims or makes the claims of an empirical science of human behavior. Rather we should view it as a branch of mathematics, one devoted to examining the formal properties of a set of assumptions about the transitivity of abstract relations: axioms that implicitly define a technical notion of "rationality," just as geometry examines the formal properties of abstract points and lines. The abstract term "rationality" may have far more potential interpretations than economists themselves realize,[7] but rather less bearing on human behavior and its consequences than we have unreasonably demanded economists reveal.

The notion that microeconomics is a branch of applied mathematics does economists more credit than several possible alternative explanations for its empirical weakness. It does not, for example, simply write off economic theory as the ideological rationalization of bourgeois capitalism; it renders the immense amount of sheer genius bestowed on the development of this theory its due. The explanation does not stigmatize the methods of economists as conceptually confused or misdirected. It isolates the limitations of the theory in a factual supposition about the determinants of human behavior, one that economists share

7. I discuss at least one of these applications, in biology, in the next section.

with all of us. But the supposition we all share is false, and so economics rests on a purely contingent, though nevertheless central, mistaken belief, just as the conviction that Euclidean geometry was the science of space rested on a purely contingent, almost equally central mistaken belief that the paths of light rays are Euclidean straight lines. Both geometry and economic theory turn out to be branches of mathematics after all.

BUT IS IT SCIENCE?

The hostility with which some economists writing on methodology greeted the claim that microeconomic theory is a branch of applied mathematics is surprising. When it was first advanced, this view was attacked at length by E. Roy Weintraub and rejected quite forcefully by Wade Hands, two distinguished economists.[8] In assessing their rejoinders below, we need to keep in mind a question they need to answer in rejecting this interpretation of economic theory. Would they rather their work were stigmatized as ideological self-deception or the result of embracing a wrong-headed epistemology or the product of a conceptual confusion?

Wade Hands's reaction to the view elaborated above was perhaps typical of what one might expect from economists eager to defend the empirical character of their discipline. Like Weintraub, Hands challenges the claim that economics is "merely mathematics," as he calls it, by citing evidence of its predictive successes. Unlike others, however, Hands ended up acknowledging that general equilibrium theory may well be a piece of applied mathematics, but nevertheless, he claims it has considerable value for social science even if it is a branch of applied mathematics.

Defending microeconomic theory by citing the alleged successes of macroeconomics, as Hands did, gives two different

8. Alexander Rosenberg, "If Economics Isn't Science, What Is It?" *Philosophical Forum* 14 (1983): 296–314; E. Roy Weintraub, *General Equilibrium Analysis: Studies in Appraisal* (Cambridge: Cambridge University Press, 1984); and Wade Hands, "What Economics Is Not: An Economist's Response to Rosenberg," *Philosophy of Science* 51 (1984): 495–503. Page references in this section are to Hand's article.

hostages to fortune. First, there is the well-known problem that macroeconomics lacks satisfactory connection to microfoundations, so that what predictive successes it has had are hard to transmit upward to any microtheory from which macroeconomic theory might be said to follow. Second, macroeconomic theory has had a relatively poor record of predictive success of its own.

But Hands's second tack, of seeking an important role for economic theory even if it is a species of applied mathematics, is a fruitful one. Never mind that it begins by admitting the very thesis of mine Hands wanted to reject. The fact is, mathematics, pure or applied, may not be "science," but from that it does not follow that it is without great import for our cognitive activities. In arguing that microeconomic theory constitutes a branch of applied mathematics, my aim was not to demean it, to cast aspersions on the subject, to stigmatize it as without import. My aim was to help us understand it and its role in scientific inquiry. Hands cites two functions for general equilibrium theory in particular:

> First, for all its inadequacies, there is a sense in which general equilibrium theory represents an apogee for economic theory. It has achieved a degree of formal rigor and sophistication comparable with the greatest physical theories, a sophistication which makes every other social science seem woefully parochial in comparison. Granted, elegance is neither necessary nor sufficient for science, but it certainly keeps economics above the muddle which often reigns in other social sciences. . . . Second, . . . the theory has provided invaluable heuristic guidance for more empirical theories. In other words the theory has *pragmatic merit.* Given the complexities of the social world, pragmatic merit may be one of the only practical criteria of theory choice. (P. 502)

Hands's first suggestion sounds like part of a problem, rather than part of its solution. After all, it is a constant point of criticism that general equilibrium theory is so abstract and out of touch with social scientific realities because it represents an attempt to ape the formalism of physical theory. What Hands and others (like Gerard Debreu) see as freedom from muddle may just as well be viewed as irrelevance. Hands's second claim is far from clear; for the notion of pragmatic merit is never explained. In fact, in my own initial suggestion that we treat microeconomic theory as a branch of applied mathematics, the aim was to exempt it from assessment against anything like a standard of pragmatic merit. Or at any rate, the view I defend is roughly that economic theory should be held up to standards of pragmatic merit no more stringently than any (other) part of mathematics.

The remarkable thing about the history of mathematics is this: though it has proceeded over centuries in accordance with an agenda of problems largely set for itself, it has remarkably enough also prefigured and provided mathematical formalism that the sciences have required just as and when needed. At any rate this appears true ancedotally and in hindsight. Perhaps it is a sufficient justification for encouraging microeconomic research to pursue its own agenda without much concern for whether it is science or not.

Indeed, treated as a body of pure research and shorn of its intended interpretation, parts of general equilibrium theory and game theory have begun to have an impact well beyond economics, in biology. Again, there is a parallel to apparently useless nineteenth-century research on non-Euclidean geometry, which in the twentieth century proved essential to the general theory of relativity. The formalism and results of general equilibrium theory are turning out to have applications for establishing stability conditions for ecosystems, applications undreamed of by the economists who proved the existence, uniqueness, and stability theorems in economics. Their results are being taken over and reinterpreted by mathematical ecology: stripped of their intentional interpretation, they provide proofs and stability conditions for unique stable equilibria that modern evolutionary biology requires in the development of its own extremal theory of balance and competition in the evolution of the biosphere.[9]

9. See, e.g., R. Oster and R. D. Wilson, *The Social Insects* (Cambridge: Harvard University Press, 1978); and R. May, *Stability and Complexity in Model Eco-*

Indeed, when recognized as a part of contractarian political philosophy, rational choice and conditions of general equilibrium have emerged as crucial tools of the sociobiologist. Sociobiology hopes to explain how norms of cooperation have emerged as stable equilibria among agents acting so as to maximize genetic fitness, or its short-sighted surrogate, ordinal preference. The possibility of free riding and the consequent failure to provide public goods at optimal levels has daunted sociobiological speculation about the emergence of culture. If the rational choice theorist can show that rational agents would choose other-regarding outcomes as maximizing ones, then both some forms of contractarianism and some forms of Darwinism about cultural evolution will have been vindicated. These results are a payoff hardly anyone could have expected. Interestingly, they reflect the compatibility of viewing economic theory both as an a priori discipline and as an exercise in contractarian political philosophy. The most powerful results to be hoped for in this area are proofs that a particular socially desirable strategy commends itself to a thoroughgoing fitness or utility maximizer. But such a proof would by no means show that the strategy did actually appear because it maximized fitness. Nothing could convert such a possibility into a confirmed account of how sociality emerged. The fossil record preserves no relevant evidence. The result is always at most the demonstration of an abstract possibility, a suggestive piece of applied mathematics.

Can we expect more of a payoff to the development of pure general equilibrium theory, or for that matter more from the derivation of theorems about strategies in parametric and strategic games? The question is no harder or easier to answer than the equivalent answer in mathematics. Can we expect payoffs in mathematics? Yes, of course we can. Can we say in advance what

---

Systems (Princeton: Princeton University Press, 1973). Game theory was introduced into evolutionary contexts as early as R. Lewontin, "Evolution and the Theory of Games," *Journal of Theoretical Biology* 1 (1961): 382–403. An excellent introduction is to be found in J. Maynard Smith, *Evolution and the Theory of Games* (Cambridge: Cambridge University Press, 1982), in which economists will recognize much of their handiwork, sometimes under different names.

they are, where they will come from, and when? Emphatically
not. Is this the slightest reason either to deny import to mathematics or to suggest that attention should not continue to be lav-
ished upon it? On the contrary, it is reason to give it a slack rein,
to give it all the intellectual freedom it demands, and to leave the
assessment of what is good and what is bad mathematics to mathematicians. Mutatis mutandis for economic theory.

But then, it will be asked, what is the cash-value difference
between this laissez-faire attitude to economics and McCloskey's
claim that what is good and bad, right and wrong, true and
false, in economics is just what pleases, convinces, and sways
economists? There are two large differences between this view
and McCloskey's, one conceptual and one practical. The conceptual difference is obvious. If economic theory is a branch of
mathematics, then it is to be held up to the standards of that discipline. And these standards are standards for knowledge; they
assume that in mathematics it is possible to acquire the truth, in
some realist's sense according to which what is true may well not
be what happens to convince mathematicians at any given moment. There is a fact of the matter in mathematics, and so too
there must be in economics, even if this fact turns out to be as
abstract in economics as it is in mathematics.

The practical difference between my view and the rhetorician's has to do with the bearing of economic theory on policy.
If economics is best viewed as more akin to a branch of mathematics on the intersection between pure axiomatization and
applied geometry, then our long-term perspective on the bearing of economic theory on policy must be qualified. And the
vacuum that economic theory leaves in the guidance of policy
must be filled by something else, something that will provide
improvable guidance to policy, both private and public. That
there might be an alternative to economic theory that is better
at policy guidance is an alternative the rhetorician declines to
face. Presumably, in the rhetorician's view, the reflexive nature
of economic theory is a sort of impossibility proof that any other
theory can do better than the poor job McCloskey admits economics does today. This pessimistic view is not one my view
need share with the rhetorical approach.

As I have noted, philosophers and economists sometimes advert to the normative interpretation of microeconomic theory. Microeconomic theories are treated as hypothetical imperatives, whose consequents enjoin action on the hypothesis that you wish to be economically rational or to encourage it in others. These theories identify incentives and disincentives for rational individuals and suggest ways of attaining, say, a government's aims by putting these incentives and disincentives into play. If the moral of my story about microeconomic theory is right, then we cannot expect the long-term sharpening of such hypothetical imperatives to provide better and better means of manipulating individual choice. In fact, the use of these hypothetical imperatives will have to be hedged around with qualification and our expectations based on them must reflect readiness to deal with completely unanticipated consequences. Not only will people fail to behave rationally, but there will be no systematic correction that we can apply to our hypothetical imperatives that will enable them to be used more reliably. And all this is because these imperatives, when treated as descriptive generalizations about what rational agents do instead of what they ought to do, are not subject to systematic improvement by the appeal to other factors that work with them to produce action.

In this respect, microeconomic theory is worse off than Euclidean geometry. Geometry too can serve as a body of hypothetical imperatives—for the use of surveyors. The paths of light rays so nearly reflect geometry's assumptions that surveyors can rely on Euclidean geometry and require no corrections. Cosmologists, however, do, and when they employ the imperatives of geometry to realize their measurement aims, they must add corrections that compensate for the effects of massive objects on light rays. Because these effects are few in number and uniform in their consequences, we can easily adjust and accommodate for them. Such adjustment and accommodation is just what we cannot do in economics. And when you add in that besides appealing to hypothetical imperatives about individual action, public policy requires also that we aggregate these imperatives to suggest governmental microeconomic policies, the problem of correction multiplies. Thus, we should neither attach unqualified confidence to predictions made on the basis of microeconomic theory nor condemn it severely when these predictions fail. Microeconomic theory can no more be faulted than Euclidean geometry should be in the context of astrophysics.

Does this attitude leave a vacuum in the foundations of public policy which will be filled by some other theory offered to serve as the basis for hypothetical imperatives useful in public policy? Perhaps before the events of the late eighties and nineties, proponents of the economic theory behind central state planning might have come forward with such a candidate. But nowadays, only the most brazen Marxian economist would do so. On the other hand, if as I argued in chapter 7, there is a connection between general equilibrium theory and contractarianism's aim of protecting all against the egoism of each, the generic powers of the theory have some bearing on policy. But it will be a relatively laissez-faire bearing. Until another theory turns up, if we insist on interpretations that accord economic theory empirical bearing on the way the world actually works, we should not apply it to justify policy beyond its generic powers to prove how the actual is possible, and how the possible might be actual. On the other hand, interpreted as a body of mathematical axioms, lemmas, and theorems, economic theory offers us even less by way of guidance.

Either way, these interpretations create a vacuum for the real guidance of public and particularly economic policy beyond the merely generic, which alternative theories will be eager to fill. To the extent that they trade in the preferences and expectations of individuals, they will do no better than neoclassical economics. To the extent that they ignore tastes and information, they will probably do worse. Accordingly, the laissez-faire economists were probably right, albeit for the wrong reasons. This is not the best of all possible worlds, but it is very easy to make it worse.

BIBLIOGRAPHY

Akerlof, George. "The Market for 'Lemons': Quality Uncertainty and the Market Mechanism." *Quarterly Journal of Economics* 84 (1970): 488–500.

Arrow, Kenneth, and Debreu, Gerard. "Existence of an Equilibrium for a Competitive Economy." *Econometrica* 22 (1954): 376–86.

Arrow, Kenneth, and Hahn, Frank. *General Competitive Analysis*. San Francisco: Holden Day, 1971.

Barrs, Robert. "Are Government Bonds Net Wealth? *Journal of Political Economy* 82 (1974): 1095–1118.

Becker, Gary. "Irrational Behavior and Economic Theory." *Journal of Political Economy* 70 (1962): 1–13.

———. *The Economic Approach to Human Behavior*. Chicago: University of Chicago Press, 1976.

———. *A Treatise on the Family*. 2d ed. Cambridge: Harvard University Press, 1991.

Bell, David, and Kristol, Irving, eds. *The Crisis in Economic Theory*. New York: Basic Books, 1981.

Blaug, Mark. *Economic Theory in Retrospect*. 3d ed. Cambridge: Cambridge University Press, 1978.

———. *The Methodology of Economics*. Cambridge: Cambridge University Press, 1980.

Booth, Wayne. *Modern Dogma and the Rhetoric of Assent*. Chicago: University of Chicago Press, 1974.

Boudon, Raymond. *A quoi sert la notion de structure?* Paris: NRF, 1968.

Braithwaite, R. B. *The Foundations of Mathematics and Other Essays*. London: Routledge and Kegan Paul, 1930.

Brown, E. K. "The Neoclassical and Post-Keynesian Research Programs." *Review of Social Economy* 39/2 (1981): 111–32.

Buchanan, James. "Constitutional Economics." In *Explorations into Constitutional Economics*, pp. 57–70. College Station: Texas A & M University Press, 1989.

Carnap, Rudolph. "The Methodological Character of Theoretical Concepts." *Minnesota Studies in the Philosophy of Science*, vol. 1, edited by Herbert Feigl and Michael Scriven, pp. 38–76. Minneapolis: University of Minnesota Press, 1956.

Cross, R. "The Duhem-Quine Thesis, Lakatos, and the Appraisal of Theories in Macroeconomics." *Economic Journal* 92 (1982): 320–40.

———. "Monetarism and Duhem's Thesis." In *Economics in Disarray,* edited by Peter Wiles and Guy Routh. Oxford: Basil Blackwell, 1984.

Debreu, Gerard. *Theory of Value.* New York: Wiley, 1959.

———. "The Mathematization of Economics." *American Economic Review* 81 (1991): 1–7.

DeMarchi, Neil. *The Popperian Legacy in Economics.* Cambridge: Cambridge University Press, 1988.

Duhem, Pierre. *Aim and Structure of Physical Theory.* Translated by P. P. Weiner. Princeton: Princeton University Press, 1954. Originally published 1904.

Einhorn, Hillel, and Hogarth, Robin. "Decision Making under Ambiguity." In *Rational Choice,* edited by R. Hogarth and M. Reder, pp. 41–66. Chicago: University of Chicago Press, 1986.

Elster, Jon. "When Rationality Fails." In *Solomonic Judgements,* pp. 1–35. Cambridge: Cambridge University Press, 1988.

Feyerabend, Paul. *Against Method.* London: New Left Books/Verso, 1975.

Friedman, Milton. *Essays in Positive Economics.* Chicago: University of Chicago Press, 1953.

———. "Theory and Realism—A Reply." *American Economic Review* 54 (1964): 736–40.

Fulton, G. "Research Programs in Economics." *History of Political Economy* 16 (1984): 187–206.

Gibbard, Alan, and Varian, Hal. "Economic Models." *Journal of Philosophy* 75 (1978): 664–77.

Goodman, Nelson. *Fact, Fiction and Forecast.* Indianapolis: Bobbs-Merrill, 1955.

Granger, Gilles Gaston. *Pensée humane et le science d'homme.* Paris: Aubier, 1983.

Hahn, Frank. *Equilibrium and Macroeconomics.* Oxford: Basil Blackwell, 1984.

Hahn, Frank, and Negishi, T. "A Theorem on Non-*Tâtonnement* Stability." *Econometrica* 30 (1962): 463–69.

Hands, D. Wade. "Methodology of Economic Research Programs." *Philosophy of Social Science* 9 (1979): 293–303.

———. "What Economics Is Not: An Economist's Response to Rosenberg." *Philosophy of Science* 51 (1984): 495–503.

———. "Second Thoughts on Lakatos." *History of Political Economy* 17 (1985): 1–16.

Hausman, Daniel. "John Stuart Mill's Philosophy of Economics." *Philosophy of Science* 48 (1981): 363–85.

———. *Capital, Prices and Profits.* New York: Columbia University Press, 1981.

———. *The Separate and Inexact Science of Economics.* Cambridge: Cambridge University Press, 1991.

Hempel, Carl. "Studies in the Logic of Confirmation." *Mind* 54 (1945): 1–26.

Hicks, Sir John. *Value and Capital*. 2d ed. Oxford: Oxford University Press, 1946.

Hobbes, Thomas. *Leviathan*. Harmondsworth, U.K.: Penguin Books, 1968.

Hogarth, Robin, and Reder, M., eds. *Rational Choice*. Chicago: University of Chicago Press, 1986.

Hollis, Martin, and Nell, Edward. *Rational Economic Man*. Cambridge: Cambridge University Press, 1975.

Hume, David. "On the Independency of Parliament." In *Essays, Moral, Political, and Literary*. Indianapolis: Liberty Classics, 1985.

Hurwicz, L. "Optimality and Informational Efficiency in Resource Allocation Processes." In *Mathematical Methods in the Social Sciences*, edited by Kenneth Arrow, Steven Karlin, and Patrick Suppes, pp. 27–46. Stanford: Stanford University Press, 1960.

Hutchison, Terence W. *The Significance and Basic Postulates of Economics*. 1938; reprint, New York: A. M. Kelley, 1960.

Kantor, B. "Rational Expectations and Economic Thought." *Journal of Economic Literature* 17 (1979): 1427–41.

Koopmans, Trilling. *Three Essays on the State of Economic Science*. New York: McGraw-Hill, 1956.

Kuhn, Thomas. *Structure of Scientific Revolutions*. Chicago: University of Chicago Press, 1962.

Lakatos, Imre. "Proofs and Refutations." *British Journal for the Philosophy of Science* 14 (1963): 1–117.

———. "The Methodology of Scientific Research Programmes." In *Philosophical Papers*, Vol. 1, pp. 1–74. Cambridge: Cambridge University Press, 1978.

Latsis, Spiro, ed. *Method and Appraisal in Economics*. Cambridge: Cambridge University Press, 1976.

Laudan, Larry. *Science and Hypothesis*. Dordrecht: Reidel, 1981.

Leontief, Wassily. "Academic Economics." *Science*, 9 July 1982, p. xii.

———. "Theoretical Assumption and Nonobservable Facts." *Essays in Economics*. New Brunswick: Transaction Books, 1985.

Lewontin, Richard. "Evolution and the Theory of Games." *Journal of Theoretical Biology* 1 (1961): 382–403.

Lucas, R. E. "Understanding Business Cycles." In *Stabilization of the Domestic and International Economy*, edited by K. Brunner and A. Meltzer, pp. 7–30. Amsterdam: North-Holland, 1977.

———. "Unemployment Policy." *American Economic Review* 68 (1978): 353–57.

McCloskey, Donald. *The Rhetoric of Economics*. Madison: University of Wisconsin Press, 1985.

———. *If You're So Smart*. Chicago: University of Chicago Press, 1990.

Machlup, Fritz. *Methodology of Economics and Other Essays*. New York: Academic Press, 1978.

Mackie, John L. *The Cement of the Universe*. Oxford: Oxford University Press, 1974.

May, Richard. *Stability and Complexity in Model Eco-Systems*. Princeton: Princeton University Press, 1973.

Mill, John Stuart. *A System of Logic*. 1949; reprint, London: Longman's, 1983.

———. *The Collected Works of John Stuart Mill*, vols. 2–4. Toronto: University of Toronto Press, 1965–67.

Morowski, Philip. *More Heat than Light*. Cambridge: Cambridge University Press, 1989.

Muth, John. "Rational Expectations and the Theory of Price Movements." *Econometrica* 29 (1961): 315–35.

Nelson, Alan. "Explanation and Justification in Political Philosophy." *Ethics* 97 (1986): 154–76.

———. "New Individualist Foundations for Economics." *Nous* 20 (1986): 469–90.

———. "Review of Hahn, Equilibrium and Macroeconomics." *Economics and Philosophy* 2 (1986): 149.

Nozick, Robert. *Anarchy, State and Utopia*. New York: Basic Books, 1974.

Oster, R., and Wilson, R. D. *The Social Insects*. Cambridge: Harvard University Press, 1978.

Quine, Willard V. O. "Two Dogmas of Empiricism." In *From a Logical Point of View*. Cambridge: Harvard University Press, 1953.

Quirk, J., and Saposnik, R. *Introduction to General Equilibrium and Welfare Economics*. New York: McGraw-Hill, 1968.

Robbins, Lionel. *An Essay on the Nature of Significance of Economic Science*. London: St. Martins, 1932.

———. "On Latsis' *Method and Appraisal in Economics:* A Review Essay." *Philosophy of Social Science* 17 (1979): 996–1004.

Rosenberg, Alexander. *Microeconomic Laws*. Pittsburgh: University of Pittsburgh Press, 1976.

———. "Can Economic Theory Explain Everything?" *Philosophy of Social Science* 9 (1979): 509–29.

———. "Obstacles to Nomological Connection of Reasons and Actions." *Philosophy of Science* 10 (1980): 79–91.

———. "A Skeptical History of Economic Theory." *Theory and Decision* 12 (1980): 79–93.

———. *Formal Thought and the Science of Man*. Dordrecht: Reidel, 1983.

———. "If Economics Isn't Science, What Is It?" *Philosophical Forum* 14 (1983): 296–314.

———. *The Structure of Biological Science*. Cambridge: Cambridge University Press, 1985.

———. "The Explanatory Role of Existence Proofs." *Ethics* 97 (1986): 177–87.

———. "What Rosenberg's Philosophy of Economics Is Not." *Philosophy of Science* 53 (1986): 127–32.

———. *The Philosophy of Social Science*. Boulder: Westview Press, 1988.

Samuelson, Paul. *Foundations of Economic Analysis*. Cambridge: MIT Press, 1947.

————. "Problems of Methodology—Discussion." *American Economic Review* 52 (1963): 232–36.

————. "Theory and Realism—A Reply." *American Economic Review* 54 (1964): 736–40.

Simon, Herbert. "Rationality in Psychology and Economics." In *Rational Choice,* edited by R. Hogarth and M. Reder, pp. 25–40. Chicago: University of Chicago Press, 1986.

Smith, Adam. "Input-Output Economics." *Scientific American* 185 (1951): 15–21.

Smith, J. Maynard. *Evolution and the Theory of Games.* Cambridge: Cambridge University Press, 1982.

Solow, Robert. "Technical Change and the Aggregate Production Function." *Review of Economics and Statistics* 39 (1957): 312–20.

Tobin, James. "How Dead Is Keynes?" *Economic Inquiry* 15 (1977): 459–68.

Tversky, Amos, and Kahneman, Daniel. "Rational Choice and the Framing of Decisions." In *Rational Choice,* edited by R. Hogarth and M. Reder, pp. 67–94. Chicago: University of Chicago Press, 1986.

von Mises, Ludwig. *Human Action.* New Haven: Yale University Press, 1948.

Wald, Abraham. "On Some Systems of Equations for Mathematical Economics." *Econometrica* 19 (1991): 368–403.

Walras, Leon. *Elements of Pure Economics.* Translated by W. Jaffe. Homewood, Ill.: Irwin, 1954.

Weintraub, E. Roy. *Microfoundations.* Cambridge: Cambridge University Press, 1979.

————. *General Equilibrium Analysis: Studies in Appraisal.* Cambridge: Cambridge University Press, 1985.

————. "Rosenberg's 'Lakatosian Consolations for Economists' Comment." *Economics and Philosophy* 3 (1987): 140.

————. *Stabilizing Dynamics: Constructing Economic Knowledge.* Cambridge: Cambridge University Press, 1991.

Wicksteed, Y. S. *The Common Sense of Political Economy.* London: Macmillan, 1910.

Wiles, P., and Routh, G., eds. *Economics in Disarray.* Oxford: Basil Blackwell, 1984.

INDEX